The Landscapes of Science and Religion

The Landscapes of Science and Religion

What Are We Disagreeing About?

NICK SPENCER
AND
HANNAH WAITE

Great Clarendon Street, Oxford, OX2 6DP,
United Kingdom

Oxford University Press is a department of the University of Oxford.
It furthers the University's objective of excellence in research, scholarship,
and education by publishing worldwide. Oxford is a registered trade mark of
Oxford University Press in the UK and in certain other countries

© Nick Spencer and Hannah Waite 2025

The moral rights of the authors have been asserted

All rights reserved. No part of this publication may be reproduced, stored in
a retrieval system, or transmitted, in any form or by any means, without the
prior permission in writing of Oxford University Press, or as expressly permitted
by law, by licence or under terms agreed with the appropriate reprographics
rights organization. Enquiries concerning reproduction outside the scope of the
above should be sent to the Rights Department, Oxford University Press, at the
address above

You must not circulate this work in any other form
and you must impose this same condition on any acquirer

Published in the United States of America by Oxford University Press
198 Madison Avenue, New York, NY 10016, United States of America

British Library Cataloguing in Publication Data

Data available

Library of Congress Control Number: 2024939874

ISBN 9780198878759

DOI: 10.1093/oso/9780198878759.001.0001

Printed and bound by
CPI Group (UK) Ltd, Croydon, CR0 4YY

Links to third party websites are provided by Oxford in good faith and
for information only. Oxford disclaims any responsibility for the materials
contained in any third party website referenced in this work.

The manufacturer's authorised representative in the EU for product safety is
Oxford University Press España S.A. of El Parque Empresarial San Fernando de Henares, Avenida
de Castilla, 2 – 28830 Madrid (www.oup.es/en or
product.safety@oup.com). OUP España S.A. also acts as importer into Spain
of products made by the manufacturer.

We dedicate this book to the memory of
Professor Tom McLeish
who was a model of 'faith and wisdom',
passion, clarity, kindness, and general all-round decency,
and who stood on the advisory group for our project.

He died in 2023 and is deeply missed by all who knew him.
May he rest in peace and rise in glory.

Acknowledgements

This book is the result of a three-year research project, which was itself the result of an 18-month scoping study. What with the time spent on funding application and on subsequent write up, the whole enterprise has taken about seven years. Over that time, we have incurred various debts that we would like to acknowledge here.

Both projects were funded by Templeton Religion Trust, and we would like to register our profound thanks for their recognition of the value of this work and their investment in the research. That research was conducted in partnership with The Faraday Institute for Science and Religion, in particular with Keith Fox and Paul Ewart, with whom it has been a pleasure to work.

Our colleagues at Theos have provided a constant source of stimulation and engagement over the years, not least as Covid rolled in just as we had started our fieldwork. That fieldwork encompassed over a hundred in-depth interviews with leading academics and communicators in various different fields of science, religion, and the communication of both. Their names are listed in appendix 1, though the order there does not correspond to the numbers identifying each interview, interviews being conducted on an anonymous basis. We are grateful to everyone we spoke to for their time and expertise.

Our project drew on an advisory group—Ben Clements, Bev Botting, Denis Alexander, Elaine Howard Ecklund, Eric Kaufmann, Fern Elsdon-Baker, Grace Davie, John Cornwell, John Hedley Brooke, and Stephen Bullivant—all of whom we want to thank. Two other members of the advisory group demand particular thanks, however.

Peter Harrison has long been a giant in the field of science and religion and his superb book on *The Territories of Science and Religion* was what really set the hares racing for us. The book is a model of erudition and scholarship, and we feel vaguely fraudulent in building on and developing Peter's controlling metaphor. We sincerely hope that our construction is, well, constructive.

It is to the final member of the advisory group, Tom McLeish, who supported the work from its inception, provided our initial introduction to OUP, and who died tragically young in 2022, to whom we owe our greatest thanks and to whom we gratefully dedicate this book.

Contents

1. Introduction: The Landscapes of Science and Religion	1
Science and religion: the relationship that will not die	1
The (shifting) territories of science and religion	3
Territories and the path away from essentialism	6
Disaggregating 'science' and 'religion'	8
The basis for disaggregation: 'science', 'religion', and meaning as usage	12
The pattern of disaggregation: 'science', 'religion', and family resemblances	14
Science and religion: from territories to landscapes	16
The foundation and structure of this book	18
Notes	22

PART I SURVEYING THE LANDSCAPES OF SCIENCE AND RELIGION: DEFINING AND DISAGGREGATING TERMS

2. Defining Science	29
The need to define science	29
The attempts to define science	33
A 'family resemblance' approach to defining science	39
The awkward reality of science	41
Conclusion: defining science	44
Notes	44
3. Disaggregating Science	47
Introduction	47
Disaggregated 'features' of science	48
The methods of science	49
The subject of science	54
The presuppositions of science	57
The objectives of science	59

viii CONTENTS

The values of science	61
The institutional structure of science	64
Conclusion: outliers and dissensions	67
Notes	69

4. Defining Religion — 73

Introduction	73
Defining religion in English law	73
Defining religion in American law	77
The polysemy of religion: anthropological, philosophical, and sociological contributions	82
Conclusion: defining religion	87
Notes	88

5. Disaggregating Religion — 90

Introduction	90
Transcendence	91
Belief	94
Commitment	98
Formalisation of belief in theology and doctrine	101
Formalisation of commitment in rituals, laws, and offices	103
Conclusion: outliers and dissensions	105
Coda: conflict by definition	107
Notes	109

6. Disaggregating Science and Religion: Public Views — 112

Introduction	112
The contours of public understanding of science and of religion	113
Religious demographics	113
Education	114
Media	115
Knowledge	117
Conclusion	119
Public understanding of (what) science (is)	120
Public understanding of (what) religion (is)	124
Conclusion	128
Notes	129

Conclusion to Part I	132

CONTENTS ix

PART II TOURING THE LANDSCAPES OF SCIENCE AND RELIGION: UNDERSTANDING WHERE THE DISAGREEMENTS LIE

Introduction to Part II	141

7. Metaphysics — 144
- Introduction — 144
- Religion's commitment to supernaturalism — 147
- Science's commitment to naturalism — 150
- The microcosm of miracles — 154
- Outstanding questions — 159
- Notes — 161

8. Methodology — 165
- Introduction — 165
- Faith as a methodological shortcut — 166
- Re-evaluating the methodological dichotomy — 169
  - Revisiting scientific methodology — 170
  - Revisiting religious methodology — 177
- Outstanding questions — 183
- Notes — 189

9. Anthropology — 195
- Introduction — 195
- Cosmology — 198
- Evolution — 202
- Anthropology — 204
- The pushback — 209
  - Revisiting the science front — 210
  - Revisiting the religion front — 213
  - Revisiting the whole encounter — 216
- Outstanding questions — 219
- Notes — 223

10. Public Authority and Public Reasoning — 227
- Introduction — 227
- Battles over abortion, and euthanasia, and genetic engineering, and ... — 231
- The difference over 'givenness' — 234

X CONTENTS

Closing the gap	237
Public authority and public reasoning	242
Outstanding questions	246
Notes	247
Conclusion	250
Appendices	254
Bibliography	260
Index	265

Unless we get inside the language games/paradigms of the respective traditions—and this necessarily requires their construction as concrete historical practices rather than as abstract philosophical traditions—the resulting conflict or consonance [between science and religion] will be a contrived artefact of the tools of analysis and translation.

John Milbank and Peter Harrison (eds.),
After Science and Religion

If you are going to disagree with your adversary in a debate in the public square, I want you to disagree with them for the right reason.

John Evans, *Morals Not Knowledge: Recasting the Contemporary U.S. Conflict between Religion and Science*

'You recognise, I hope, the existence of spiritual interests in your patients?'

'Certainly I do. But those words are apt to cover different meanings to different minds.'

Mr Bulstrode and Dr Lydgate in conversation
George Elliot, *Middlemarch*

1

Introduction

The Landscapes of Science and Religion

Science and religion: the relationship that will not die

The relationship between science and religion is of seemingly endless fascination. At the time of drafting this opening paragraph, a Google news search for the two terms revealed recent stories about how an important paper in the scientific journal *Nature*, on the relationship between 'moralising gods' and the evolution of complex societies, had been retracted;[1] how the recently deceased cosmologist and Nobel Laureate Steven Weinberg had been famed for saying that science had proven the universe to be pointless and that 'science and religion [have] nothing constructive to say to one another';[2] how, by using datasets from patients with brain lesions, researchers had managed to trace 'locations associated with spiritual and religious belief to a specific human brain circuit';[3] and how Orthodox priests in Greece were leading the charge against Covid vaccinations[4]—to name only four.

The interest goes beyond these transient media stories. The language of the divine—*The God Equation*, *The Mind of God*, *The Serpent's Promise*—is often appropriated for popular science books, whilst the trope of scientific evidence—*Proof of Heaven: A Neurosurgeon's Journey into the Afterlife*, *The Language of God: A Scientist Presents Evidence for Belief*, *Finding God in the Waves: How I Lost My Faith and Found It Again Through Science*—is deployed for religious ends. According to Google Ngram Viewer, mention of 'science and religion' in books has risen constantly since the mid-1970s, except for a brief fall after a peak in 2006–07, since when it has been recovered. It now stands at an unprecedentedly high level.[5]

This breadth of interest is not, of course, burdened by any consensus. Indeed, the debate around science and religion is notorious not only for its persistence and popularity but also for the strength of the differing opinions it attracts. Conflict, concord, and consilience are all expressed with confidence that commonly spills over into hubris. Jerry Coyne,[6] Yves

2 THE LANDSCAPES OF SCIENCE AND RELIGION

Gingras,[7] Richard Dawkins,[8] and Victor Stenger,[9] among others, have shown themselves to be forceful, articulate, and popular advocates for the view either that science has disproven God or that science and religion are engaged in 'an impossible dialogue' or that they are straightforwardly 'incompatible'. Alister McGrath,[10] Francis Collins,[11] John Polkinghorne,[12] Tom McLeish,[13] and Celia Deane Drummond,[14] among others, have argued for a more compatibilist, indeed harmonious and mutually productive, relationship between the two. In contrast to both these positions, Stephen Jay Gould advocated the principle of 'non-overlapping magisteria' (NOMA), in the sense that science dealt with the empirical realm and religion with that of meaning and value, the two thereby existing in harmony purchased at the expense of any interaction.[15] More recently, the psychiatrist and intellectual historian Iain McGilchrist has built on his analysis of the hemispheric nature of the brain[16] to develop a sophisticated approach of 'scientific' and 'religious' understandings of reality, an approach best categorised as complementary rather than conflictual, concordant, or non-overlapping.[17]

In the light of these differing views—only a handful of which have been mentioned here—and the on-going public interest in science and religion, scholars have attempted to delineate a typology of the various positions held. Perhaps the best known of these is the four-fold categorisation proposed by Ian Barbour in *When Science Meets Religion*, according to which, science and religion can be in conflict, independent of one another, in dialogue, or fully integrated.[18] The American theologian John Haught had already suggested a similar, if more alliterative typology of conflict, contrast, contact, and confirmation.[19] In an article on 'Conflict and compatibility', theologian Ian McFarland developed Barbour's four categories by delineating two different approaches to compatibility and to integration between science and religion.[20] In an important, if somewhat neglected book on *How to Relate Science and Religion*, philosopher Mikael Stenmark outlined 'four different levels or dimensions of science and religion'.[21] Others have multiplied categories. Arguing that 'science and religion interact at so many different junctures and in so many different ways that any simple generalization is misleading', Ted Peters put forward no fewer than ten conceptual models for relating the two.[22] In the light of this profusion of positions and typologies, it is no surprise that sociologists John and Michael Evans have claimed that 'the field of religion and science is one of the muddiest in all of sociology'.[23]

One reason for this, they claim, lies 'in the long-running academic assumption that religion and science always conflict and that they conflict over competing truth claims about the world'. This might well be true, but it is the contention of this book that the problem lies deeper still. The four news items cited in the opening paragraph of this section illustrate the issue. Under the general rubric of 'science and religion', these stories cover questions of human evolution, civilisational development, morality, purpose, neuroscience, spiritual belief and experiences, public health, political action, and religious activism. This is a mixed bag by anyone's reckoning, and it underlines how speaking of science or religion in clearly delineated terms is extremely difficult, and speaking of science *and* religion in such terms even more so.

In essence, many of the debates around science and religion remain mired in essentialist concepts of the two entities—the conviction that there is such a *thing* as science and such a *thing* as religion, and that the two can be meaningfully compared with one another. Without addressing this definitional problem, we will make little progress in clarifying, let alone settling, this debate. To have a compelling view of the relationship between science and religion, we must, first, tackle the difficult issue of what exactly we are being asked to relate to one another. What, in other words, are we even talking about? In doing this, we can learn from history.

The (shifting) territories of science and religion

For over 30 years, historians have been dismantling the simplistic narrative of a relentless and longstanding 'conflict' or 'warfare' between science and religion. First outlined in detail in the final decades of the nineteenth century in the work of John William Draper and Andrew Dickson White, and then popularised by Edward Youmans in his *Popular Science* magazine, the metaphor of conflict retains significant power in the popular imagination today.[24]

Dating from the publication of either *God and Nature*, a volume of essays edited by David Lindberg and Ronald Numbers in 1986, or John Hedley Brooke's *Science and Religion: Some Historical Perspectives* in 1991, a new movement in historical scholarship—exemplified by the work of Lindberg, Numbers, and Brooke as well as that of Peter Harrison, Bernard Lightman, Geoffrey Cantor, and Edward J. Larson, among others—has successfully overturned the exhausted historical conflict narrative within academic circles.[25] Historians, most recently James Ungureanu, now study Draper and White themselves, rather than the history they purported to reveal, in order to trace the origins of the conflict narrative.[26]

4 THE LANDSCAPES OF SCIENCE AND RELIGION

For all its success, however, this scholarship has been less successful either in offering an alternative controlling metaphor with which to replace Draper and White's 'conflict' and 'warfare', or in changing wider public or elite opinion.[27] As the 2018 volume of essays on *The Warfare between Science and Religion: The Idea that Wouldn't Die* put it, 'notwithstanding all the outstanding work by a generation of historians dismantling the "conflict model", their revisionist accounts have scarcely made a dent on leading public intellectuals'.[28]

These two points—the failure of contemporary scholarship to come up with a new controlling metaphor to describe the historic (and contemporary) relationship between science and religion, *and* the on-going power of the 'conflict' metaphor in the public imagination—are almost certainly linked. What the warfare narrative lacked in historical nuance or balance, it made up for in clarity and memorability. The most widely discussed replacement for conflict is 'complexity', an idea most associated with Brooke's *Science and Religion*, although present also in *God and Nature*.[29] The problem, however, is that—in stark contrast with the Draper/White warfare metaphor—this 'complexifying history', in Numbers' phrase, 'seems to have little to recommend it besides its truth'.[30] As Brooke himself has written, 'complexity is a historical reality, not a thesis', still less a controlling metaphor.[31]

Over recent years, there have been various attempts to resolve, or at least to address this problem, by capturing the nuance and subtlety of the (historical) relationship between science and religion without altogether abandoning any attempts to draw conceptual clarity between the two. Thus, in his essay on 'Simplifying Complexity' in the *festschrift* volume *Science and Religion: New Historical Perspectives*, Numbers outlined five 'mid-scale patterns'—of naturalisation, privatisation, secularisation, globalisation, and radicalisation—which, while in no way attempting to resurrect an 'uncomplicated master-narrative', sought to establish identifiable sub-narratives amidst the complexified rubble of the ruined warfare narrative. More recently, Bernie Lightman edited a volume of essays on *Rethinking History, Science and Religion* which evaluates the use of complexity as a heuristic principle by paying particular attention to spaces, geographical and media, in which the relationship between science and religion has been mediated. Understanding science and religion *spatially* is one possible way of maintaining complexity with a measure of conceptual clarity.[32]

In this vein and most influentially, Peter Harrison has sought to combine the reality of historical nuance and complexity within a coherent and instructive spatial framework through his influential metaphor of 'territories'. Initially in his 2006 essay on 'Constructing the boundaries' between science and religion, and then, more substantially, in his 2015 book *The Territories of Science and Religion*, Harrison brought considerable historical scholarship to bear on the argument that intellectual territories, like national ones, shift over time, and that it is only by paying attention to the those changes that the historical relationship between entities like 'science' and 'religion' can be determined.[33]

Neither 'science' nor 'religion' is self-evident or a 'natural' category. On the contrary, both are socially constructed and have shifted considerably in their meanings over the years. Natural philosophy was, in the ancient Mediterranean world, the study of the natural world as a means to achieving the good life. Mathematics and physics were similarly ethical-religious activities, particularly according to neo-Platonic understanding. *Scientia*, in the Middle Ages, was a mental habit oriented to the acquisition of knowledge through the 'rehearsal of logical demonstrations'. *Scientiae* were the different bodies of knowledge derived from the exercise of this habit. Over time, such diverse virtues, practices, and bodies of knowledge coalesced into an allegedly unitary 'method', supposedly free of any metaphysical or existential baggage, by means of which 'nature', another constructed and contested concept, could be successfully interrogated.[34]

It was a similar story for 'religion'. *Religio* was once the scholastic virtue of inner piety and was transformed, via the confessionalisation of Europe and the European encounter with other global cultures, into a set of purportedly coherent and formalised beliefs, doctrines, and practices, through the lens of which, first Christian denominations and then 'world religions' were identified and essentialised. Historically, 'religion' has as much (or as little) coherence or constancy as 'science'. In this way, any straightforward relationship between 'science' and 'religion' in any period before the mid-nineteenth century, whether conflictual or harmonious, would have been inaccurate, to the point of meaningless.[35]

Harrison's territorial metaphor has significant potential. Recognising the changing nature of, and borders between, these intellectual territories helps rescue the historical interpretation of science and religion from the Scylla of simplicity and the Charybdis of confusing complexity. Territories is a perceptive and useful metaphor for understanding the relationship between 'science' and 'religion'.

The concept of intellectual cartography is not, of course, a new one, and it has been deployed before in this field. The philosopher Mary

6 THE LANDSCAPES OF SCIENCE AND RELIGION

Midgley, for example, drew on the idea of intellectual maps in her attempt to 'end the apartheid' of mind and body separation.[36] However, Harrison's sustained and highly evidenced essay has breathed fresh life into metaphorical 'cartography' as a coherent yet flexible model by means of which the relationship between 'science' and 'religion' may be understood.

Territories and the path away from essentialism

Harrison's analysis is historic and ends in the later nineteenth century when recognisably modern definitions of science and religion had been established. However, that is not the end of the story. Just as there is breadth and fluidity to the historical concepts of 'science' and of 'religion', as Harrison has shown, so there is to contemporary definitions of the words, no matter how confidently we use them today. Current debates around science and religion are as afflicted by terminological confusion and, in particular, unwarranted essentialism as historical accounts have been.

According to such essentialising logic, 'religion' and 'science' today are clearly and objectively demarcated entities, or at least sufficiently so for the relationship between them to be described with confidence. Such an approach seems to be a prerequisite for any strongly conflictual understandings of the relationship between the two. To prove that science and religion are at war, it is crucial to demonstrate that each is a coherent, identifiable, and clearly structured entity, ideally something that is defined precisely against the other. In Steven Pinker's words, 'anyone who engages in secular reason is a kind of honorary scientist ... I define science in the broadest term of relying on logic and evidence rather than on dogma or authority or subjective feeling'.[37]

That said, essentialism need not lead to conflict. Steven Jay Gould's concept of science and religion as 'non-overlapping magisteria' depends on equally essentialist premises while proposing a far more ameliorative relationship between the two than is afforded in the work of Dawkins, Coyne, or Pinker. Gould believed that (the magisterium of) science covered the empirical realm—'what the universe is made of (fact) and why does it work this way (theory)'—whereas (the magisterium of) religion extended 'over questions of ultimate meaning and moral value'.[38] His two magisteria exist as coherent bodies of thought but are wholly distinct from, rather than antagonistic towards, one another.

Moreover, an essentialist approach to the relationship between science and religion need not arrive at any single model of relationship between them, whether conflictual or non-overlapping. Indeed, the very task of developing a typology of science and religion, such as those of Barbour, Haught, and Peters mentioned above, implicitly draws on essentialist ideas of science and religion. The problem, therefore, is not that an essentialist understanding of science and religion leads necessarily to antagonism, to harmony, or to any particular relationship between the two. It is, more simply, that such an understanding is ultimately intellectually and sociologically indefensible. 'Science' and 'religion' are not single, coherent entities today any more than they were over history.

This is increasingly recognised in the scholarship on the interaction between the two. In his article 'Whose science and whose religion?' Stuart Glennan contended that those arguments about the relationship between science and religion that proceed on the basis of identifying a set of essential characteristics of scientific and religious worldviews are 'doomed to failure because science, to some extent, and religion, to a much larger extent, are cultural phenomena that are too diverse in their expressions to be characterized in terms of a unified worldview'.[39] McGrath has explained how attempts to define science and religion invariably founder on the now widespread recognition that both are 'theoretical constructs ... shaped by cultural perceptions and agendas' in a way that severely impedes 'any attempt to offer essentialist accounts of their nature, or their possible interactions'.[40] Joseph Baker, writing on 'Public perceptions of incompatibility between "science and religion"', has said that any perceived relationship between science and religion is necessarily 'socially constructed', and marked by perceptions that 'are pragmatic, heuristic, and vary depending on how individuals engage both institutions, rather than being indicative of an inherent relationship between the two'.[41] Jonathan Jong has argued that 'religion' and 'nonreligion' are 'fuzzy categories', comprising a loose collection of 'causally and phenomenologically distinct phenomena', such as 'belief in supernatural agents, participation in rituals, formation of non-kin groups, [and] obedience to moral codes', and as such 'they have no legitimate scientific use'.[42]

In short, for all that the territories of 'science' and 'religion' may have settled down after centuries of fluctuation and dispute, they remain unclear, disputable, and socially constructed. Unlike political territories, at least of the modern era, there is no sovereign political authority to determine where the boundaries lie.

8 THE LANDSCAPES OF SCIENCE AND RELIGION

Disaggregating 'science' and 'religion'

This challenge is not unique to the discourse around the relationship between science and religion. Indeed, it is particularly live and acute in political and legal theory, which wrestles with the intractable question of how states should legislate, and how courts should adjudicate, on issues of the freedom of speech, dress, association, organisation, and governance of religious groups while maintaining just and reasonable equivalence with non-religious worldviews. Faced with this complex and ambiguous understanding of religion in legal and political thought, Cécile Laborde, professor of political theory at the University of Oxford, has, in her recent book, *Liberalism's Religion*, suggested a way forward that navigates the problem of the socially constructed nature of religion by disaggregating 'religion' into identifiable elements or dimensions. On the premise that 'there is no stable, universally valid empirical referent for the category of religion', she argues that political theorists should disaggregate religion into a 'plurality of *relevant normative dimensions*', as a way of developing a more robust and equitable theory of liberal egalitarianism.[43]

This enables her to dispense with the 'Western-, Christian-inflected construal of religion' on which liberal political theory has historically relied (the development of which Harrison traces in *Territories*) and to go beyond the imprecision of the 'loose analogue of "conception of the good"' that John Rawls popularised in Anglo-American analytical political philosophy.[44] The resulting disaggregated 'dimensions' of religion include: a *cognitive dimension*, in which religiosity is predicated on certain specific beliefs about God and the nature of humans; an *identificatory or symbolic dimension*, in which certain texts, symbols, aspects of dress, etc. denote religious identity and belonging; an *ethical dimension*, in which the religious hold moral commitments that inform their conscience and structure and orient their moral lives; a *community dimension*, in which religiosity is expressed through patterns of collective worship and organisation; and a *public dimension*, in which religious commitment is manifest in modes of association from the formal and personal (like marriage) to the informal and communal (like voluntary social action) each of which, in different ways, contributes to the wider public good. By Laborde's reasoning, it is only by disaggregating religion in this way, and by identifying corresponding dimensions with similar elements of non-religiosity (personal beliefs and experiences, secular conscience, associations, comprehensive worldviews, etc.), that liberal egalitarianism can hope to treat religion with the justice and fairness that it claims.

Although not without criticism,[45] Laborde's approach to religion has been well received, and although it may at first seem as if science doesn't lend itself to the same approach—Glennan, for example, claimed that science is only 'to some extent' a cultural phenomenon—there is in fact similar potential (and need) for disaggregation here. Although (arguably) less socially constructed than 'religion', science is still a broad, 'fuzzy', constructed, and contestable category that is in need of disaggregation if we are to talk about its relationship with religion.

Such a disaggregative approach has already been ventured, albeit in an ad hoc way, in several rather different approaches within the discourse of science and religion. Alvin Plantinga's book *Where the Conflict Really Lies: Science, Religion and Naturalism* argues that simply talking of science and/or/versus religion makes little sense as it misses the crucial details. He contends that the true relationship between science and religion is a *mixture* of concord and conflict depending on which dimensions of science or religion are under scrutiny.[46] By Plantinga's reckoning, 'there is superficial conflict but deep concord between science and theistic religion, but superficial concord and deep conflict between science and the naturalism that many scientists believe is necessary to science'.[47] More precisely, there are some areas of only apparent conflict (he cites *evolution* and *miracles*); some where there is real but superficial conflict (such as evolutionary psychology); some where there is deep concord (such as the apparent fine-tuning of the universe, the unreasonable effectiveness of mathematics, or the remarkably effective match between human cognitive powers and the wider world); and some where there is deep and serious conflict (though he argues that this is not between religion and science but between religion and scientific naturalism). The relationship between science and religion is thus complexified according to the dimensions that emerge from the disaggregation of each.

In a similar vein, John Evans, Professor of Sociology at University of California (UC) San Diego, in his 2018 book *Morals Not Knowledge: Recasting the Contemporary U.S. Conflict between Religion and Science*, also seeks to dislodge the idea that there is 'a foundational conflict' between science and religion in public by emphasising the 'locatedness' of interaction, and stressing how science and religion discourse is conducted by *people*, who are always specifically located and shaped by particular concerns and objectives, rather than in the abstract.[48] Evans teases apart 'elite' discourse on science and religion, with its focus on epistemology and the nature of reality, from popular discourse, which has a much greater interest in existential, ethical, and in particular medical dimensions. The nature of the interaction between science and religion is predicated on whether it is located at the (elite) level

10 THE LANDSCAPES OF SCIENCE AND RELIGION

of truth claims or the (popular) level of moral or existential claims. A similar point is made by Stephen T. Asma with regard to the *affective* dimension of religion, which is all too often ignored in science and religion discourse.[49]

More recently, Fern Elsdon-Baker and colleagues in the *Science and Religion: Exploring the Spectrum* project have drawn out of their research the social and cultural factors that drive the narrative (and, in particular, the conflict narrative) between science and religion. They have shown how *both* religion and science 'act as a facet of social or cultural identity', and while no-one would have thought otherwise of religion, the way in which Elsdon-Baker et al. have shown how enthusiastic promotion of science, and especially evolution, can serve as 'an important role in the expression of atheistic identities' is a helpful counterweight to the sometimes implicit view that only evolution rejection serves as an identity marker. Some people 'engage with science not merely as a neutral body of knowledge but as a component of social identity' which orients them towards a particular (combative) approach to science and religion. In this way, science is disambiguated to draw out its latent social and cultural dimensions, all too often ignored in standard accounts of its relationship with religion.[50]

One of the most worked-through 'disambiguating' approaches to science and religion can be found in Mikael Stenmark's aforementioned *How to Relate Science and Religion*. Stenmark rejects the idea that science and religion can be accurately related to each other in any straightforward or one-dimensional way precisely because each entity is 'multi-dimensional'. In the light of this, he draws out four comparable dimensions. Firstly, there is a *social dimension* in which science and religion are understood and compared as 'complex and fairly coherent socially established cooperative human activities through which their practitioners ... try to achieve certain goals by means of particular strategies'. Second, there is a *teleological dimension*, by means of which science and religion are identified by their particular goals, which are further disambiguated into 'epistemic and practical, ... individual and collective goals'. Third, there is the *epistemological dimension*, in which the concerns for rationality, justification, knowledge, and truth are compared. And finally, there is the *theoretical dimension*, which encompasses the (comparably well-established) terrain of the subject matters of science and religion. Within all this, Stenmark further disaggregates science to distinguish between the 'problem-stating phase' (what he calls science$_1$), the 'development phase' (science$_2$), the 'justification phase' (science$_3$), and the 'application phrase' (science$_4$). Stenmark's approach constitutes the most comprehensive analysis of science and religion to adopt the 'disaggregative' approach, and foreshadows a number of themes developed in this book.[51]

Disaggregating 'religion' and 'science' as a means of clarifying the relationship between the two thus has a number of promising precedents. It offers a kind of contemporary intellectual cartography that would carry on from Harrison's historical cartography. Disaggregating the terms shifts the conversation away from generic discussion of science and religion (whether antagonistic like Coyne's, or ameliorative like Gould's, or varied like Barbour's) and enables us to understand what people (and indeed which people) are dis/agreeing about which dimensions of science and religion and why.

There remain two outstanding problems to this approach, however. First, just because we recognise that the terms science and religion need to be disaggregated in order to have a more accurate and fruitful discussion between the two, it is not obvious *whose* disaggregations we should favour. There are many on offer. We might select the different dimensions of religion identified by Cecile Laborde, or by Jonathan Jong, or we might favour Ninian Smart's influential six (latterly seven) dimensions, or the more straightforward division between substantive or functional elements identified by Peter Berger and others.[52] Who is to say what elements define religion—or indeed science?[53]

Second, even were we to disaggregate science and religion into an agreed number and range of constituent elements or dimensions, it is far from clear which and how many would be needed in order to recognise an entity as being 'religion' or 'science'. Jonathan Jong makes this point with regard to religion by means of what he calls the Buddhism problem and the football problem.[54] Were we, for the sake of argument, to settle on Ninian Smart's dimensions of religion (which we will mention in more detail in chapter 4), should we expect an activity to reflect all seven dimensions[55] to merit being called a religion (in which case, certain branches of Buddhism might find themselves excluded)? Or should we require only one or two dimensions (in which case, football—or indeed any number of comparable ritual, narrative, and social commitments—would be included)?[56]

These are two stubborn problems—what is the basis for disaggregation and what is the proper pattern of disaggregation?—that we need to address before we adopt this approach as a way of bringing accuracy and nuance to the current debates concerning science and religion. In answering them, we will draw on two key ideas from the later philosophy of Ludwig Wittgenstein.

12 THE LANDSCAPES OF SCIENCE AND RELIGION

The basis for disaggregation: 'science', 'religion', and meaning as usage

Wittgenstein has long been deployed in debates around science and religion, usually in an attempt to bring some clarity to the discussion or, perhaps, to transform the discussion altogether by 'show[ing] the fly the way out of the fly bottle', as he put it in *Philosophical Investigations*.[57]

Tellingly, Peter Harrison, Alister McGrath, and Cecile Laborde all (briefly) situate their conceptualisation of religion (and science) within a Wittgensteinian framework. Harrison sees his terminological excavation as analogous to Wittgenstein's conviction that philosophical problems are the result of, and can be addressed by, the untangling of misconceived and mis-used language. 'At least some of our contemporary quandaries to do with competing religious truth claims and conflicts between scientific and reli-gious doctrine arise out of the way in which we presently deploy the terms "religion" and "science"'.[58] McGrath notes early on in *Territories of Human Reason* that Wittgenstein's famous declaration that 'a *picture* held us captive' has 'particular force in the field of science and religion' where 'the culturally regnant "picture" [is] of ... perennial and essential conflict'.[59] And Laborde comments that Wittgenstein's observations about games exhibiting 'family resemblances' can be fruitfully applied to discussions about religion.[60]

This final point is, we will argue, critical to resolving the second of the challenges outlined above—namely the question of how many and which elements of the disaggregated terms are necessary for those terms to remain within the basic category of 'religion' or 'science'. In effect, how 'religious' or how 'scientific' does something have to be to merit being called 'religion' or 'science'? Before we come to that argument from *Philosophical Investi-gations*, however, we argue here that Wittgenstein's later theory of language can also help address the first problem, namely which or whose examples of disambiguation should we favour in our attempt to understand science and religion.

In contrast with his earlier, more 'atomic' understanding of language, in which language is isomorphic with the logical structure of the world, the Wittgenstein of *Philosophical Investigations* advocates an understanding that locates meaning in usage. He rejects the idea that 'the words in language name objects', and that 'sentences are combinations of such names'.[61] The meaning of a word lies not in the object or fact that it represents but in the (sometimes rather different) ways it is used. As he says in section 43, for most if not all words, 'the meaning ... is its use in the language'.

Wittgenstein illustrates this point with the example of the word 'game'. Required to define the word in an ostensive way, we soon run into trouble. There are board games, card games, ball games, Olympic games, and so on. 'What is common to them all?' The answer is nothing. Various different games share various different qualities with one other—some are competitive, some communal, some amusing, etc.—but there is no single quality that they all possess by dint of which each can be called a 'game'. There is no one common quality that denotes game-ness.[62]

In spite of this apparent confusion, people still use the word game and understand each other when they do. The meaning comes not from any discrete element or behaviour that the word clearly represents, but from the way in which people use the word in conversation with one another. Importantly, however, that understanding is predicated on shared patterns of usage and a shared wider context—or, to use Wittgenstein's preferred terms, on 'language games' and 'forms of life'. 'The speaking of language', he writes in section 23, 'is part of an activity, or of a form of life', before proceeding to offer a long list of resulting 'language games'.[63]

This understanding of language orients us in a very important direction when it comes to the task of disaggregating the meanings of science and religion. The idea of meaning as usage requires that we observe—or more precisely listen to—the way the words are being used as a way of capturing their various meanings. Wittgenstein stresses this point. 'The meaning of a name is sometimes explained by pointing to its bearer', he says in section 43, and again, in section 6 when talking about how we should define the word 'game', 'Don't think, but look!' In effect, and picking up on Harrison's remark above, listening carefully to the way we deploy the terms 'religion' and 'science' in discussion constitutes the first step in the task of clearing up some of our contemporary quandaries in the discussion around science and religion.

We can refine this point further. Listening to the ways in which the terms 'religion' and 'science' are used, as a way of clarifying and disaggregating their meanings, should entail listening to the way that they are used by a wide range of people and in a wide range of contexts, as Wittgenstein's list of games in section 66 suggests. This list includes board games like chess and children's games like 'ring-a-ring-a-roses', as well as sports games like tennis and those played in the Olympic games. Were we to listen only to what sportsmen and women consider to be a game, we might end up with an unduly circumscribed notion, or set of notions, as to what a game is. Because the use of words determines their meanings, and use can occur within a wide

14 THE LANDSCAPES OF SCIENCE AND RELIGION

range of 'language games' and 'forms of life', it is therefore important to listen widely and not pre-emptively to close down the possible range of usages (and therefore meanings).

This means two things for the study of science and religion. Firstly, taking a cue from John Evans' distinction between elite and popular discourse, it means assessing how 'experts'—such as philosophers of science, philosophers of religion, theologians, and practising scientists themselves—use the terms *but also* how 'non-experts'—from a representative range of demographic, religious, ethnic, and educational backgrounds—do so. We may want to pay more attention to those who use the words frequently and in the context of their considerable personal and professional expertise; these people should, at least in theory, have thought carefully about what they are talking about when they talk about science, or religion, or both. But we should not do so to the complete exclusion of those who may not be professionally or personally connected with religious or scientific institutions but who still know of and talk about science and religion. Usage does not simply mean usage by those who work professionally with the terms.

Secondly, it means listening to discourse around science and religion in as generous a way as possible, rather than narrowly examining the use of the terms in relation to the obvious and familiar subjects in this discussion (evolution, cosmology, etc.). In effect it means opening up the discourse to include philosophical, ethical, social, legal, and existential perspectives. Again, as Evans notes regarding the US, both science and religion play significant roles in public discourse in ethical, and in particular medical ethical, discussions. At a popular level at least, the discourse (and in particular the disagreement) between science and religion is often about morally inflected ideas concerning the nature, behaviour, and treatment of human beings, and yet these have often been ignored or downplayed in the academic discussion of science and religion. A Wittgensteinian approach to science and religion, grounded in the idea of meaning as usage, will require not only that we listen to how the terms are used, but also that we listen *widely*.

The pattern of disaggregation: 'science', 'religion', and family resemblances

Wittgenstein's idea of meaning as usage helps address the challenge of 'which or whose disaggregation of "science" and "religion" should we draw on?' It does so by pointing us towards the various dimensions of 'religion' and 'science' that emerge from listening to what (a range of) people say about them. That will naturally incorporate the disaggregations proposed by other scholars of religion, but it will also encompass a wider range of elite and popular conceptualisations that have not necessarily been consigned to academic paper(s).

There remains, however, a second challenge—namely what number or range of the disaggregated dimensions that emerge from this process are necessary or sufficient for something to be judged 'religion' or 'science'? If, for example, listening to discourse on science resulted in finding 'hypothesis', 'observation', 'inductive reasoning', and 'critical analysis' as among the disaggregated dimensions of science (as presumably it would), could an activity that exhibited these four dimensions without any others (e.g. experimentation or peer review) fairly claim to be called science? Similarly, if listening to discourse on religion resulted in finding 'deeply held personal belief', 'concerning a weighty and substantial aspect of human life', 'with serious attendant ethical demands' as among the disaggregated dimensions of religion, would a practice that exhibited only these three (but not, say, belief in supernatural entities) count as a religion?[64] The answers to these questions would determine whether activities like detective work, in the first example, and ethical veganism, in the second, could be included, respectively, as 'science' or as 'religion'.

Here, as Laborde notes, Wittgenstein's notion of 'family resemblances' is helpful. After his description of various games in section 66 of *Philosophical Investigations*, Wittgenstein concludes by remarking that the games he has discussed comprise 'a complicated network of similarities overlapping and criss-crossing', and, in the following section, he says the best way he can find to characterise this pattern of similarities and dissimilarities is the idea of 'family resemblances'.

By this analogy, members of a family will resemble one another in terms of their 'build, features, colour of eyes, gait, temperament, etc. etc.'. No-one will exactly resemble another family member, and no-one will share no common features. At the same time, there is no single feature that all family members possess that could be judged to be the essential and ubiquitous family feature. Identity does not reside in any single feature or, in the case of science and religion, any one single disaggregated dimension or constituent element.

This means that it is impossible to say definitively whether something is (in Wittgenstein's illustration) a game—a point that he readily acknowledges in section 68, when he asks himself 'What still counts as a game and what

16   THE LANDSCAPES OF SCIENCE AND RELIGION

no longer does? *Can you give the boundary?*' and answers simply 'No'.[65] A 'family resemblance' approach to language precludes making definitive definitions. It avoids essentialism.

The implications of this approach for the task of disaggregating the terms 'science' and 'religion' will be clear, but may also be disquieting. For, as Wittgenstein goes on to remark, in his interlocutor's voice, it means that 'the use of the word is unregulated ... [and] if the concept "game" is uncircumscribed like that, you don't really know what you mean by a "game"'.[66]

Superficially, this is indeed a challenge, threatening to obscure any discourse around science and religion by 'unregulating' and 'uncircumscribing' the terms altogether. However, the approach is not as problematic as all that because, as Wittgenstein says, just because we have not historically been able draw a 'boundary' around the meaning of 'game', this has 'never troubled you before when you used the word "game"'.[67] Meaning, after all, resides in usage, and just as people—indeed even children, as Wittgenstein points out—use the word 'game' without being able to define or regulate its many possible meanings, so people continue to use the words 'science' and 'religion' even though they are ultimately unregulated and uncircumscribed.

The manner in which Wittgenstein uses the word 'boundary', and discusses the lack thereof in explaining his concept of 'family resemblances', offers a neat link to Harrison's own metaphor of the territories of science and religion. Historically, the territories have moved considerably, and it has been very hard to state definitively where the boundaries around each term have lain, and impossible to delineate any permanent boundaries. However much the terms may have stabilised by the end of the nineteenth century, the nature of language, as it is illuminated by Wittgenstein's later philosophy, is such that the terms remain 'unregulated', 'uncircumscribed', and without definitive boundaries. In effect, the task of describing the indeterminate territories of science and religion continues.

Science and religion: from territories to landscapes

This is the source of the shift in metaphor from 'territories' to 'landscapes' proposed in this book. The reasons pertain to the two Wittgensteinian moves outlined above.

Firstly, 'territories', like maps, are, so to speak, seen from above. The word 'territory' is commonly used in conjunction with reference to the defined jurisdiction of an authority, or even, such as in Canada and Australia, an officially recognised division of land administered by a federal government. The attendant adjective 'territorial' often has connotations of antagonism or at least defensiveness. Territorial animals are those that reside in a particular location and are willing to protect that residence, with force if necessary.

A shift to 'language as usage' encourages us to eschew a bird's-eye, objective, and authoritative view of the terrain we are exploring. If the words 'science' and 'religion' cannot be authoritatively defined 'from above' but only understood fully by 'looking' at the way they are used, we need to define the terrain *from the ground level* of usage. Landscapes are territories seen from the ground. Understanding the *landscapes* of science and religion nudges us away from any latent idea that what we find is formerly recognised or defined by an authority. Talking of 'landscapes' of religion and science places us *within* those territories, at the level of usage, and it also forces upon us the discomfiting recognition that we cannot but view those landscapes from where we are standing, which may not be where others are standing.

The first reason for the shift in metaphor to 'landscapes' is thus grounded in Wittgenstein's understanding of language as usage. The second is grounded in his idea of 'family resemblances' and what that does to hopes of defining words unambiguously or definitively.

Territories, again like maps, are commonly seen with lines and borders and boundaries—sometimes shifting and sometimes disputed, but nonetheless 'real'. It is an adjunct to the idea of territories as formally delineated. Landscapes, by contrast, are populated by features rather than boundaries. A landscape does not have clear or authoritative markers stretching across it, dividing it from other landscapes. In a similar way, referencing Wittgenstein's rhetorical question in section 68, words and classifications do not have boundaries round them. Rather, they have various, complex, and often disordered features that characterise them. In this way, while it is difficult to draw definitive boundaries that separate science from all other activities that are not science (and, similarly, religion from all other activities that are not religion), it is possible to identify more or fewer features within the activity that mark it out as more or less 'science' (or religion).

Thus, an activity that exhibits more of the elements of, say, the Science Council's definition of science (see chapter 2)—such as what goes on in CERN or the Babraham Institute—may be more clearly and justifiably defined as 'science' than another 'scientific' activity—such as psychoanalysis or fieldwork anthropology—which, in turn, is more defensibly

18 THE LANDSCAPES OF SCIENCE AND RELIGION

termed science than what goes on in an economics seminar or a detective's notebook. The stark territorial division between 'science' and 'not-science' has been replaced, through Wittgenstein's idea of family resemblances, with a more gradated landscape, some parts of which clearly demonstrate more scientific 'features' than others.

In a similar way, when it comes to thinking about religion, there are some activities within the overall landscape, such as the Latin Mass, the Hajj, and Kumbh Mela, that are densely populated with religious features, and others, such as leaving votive candles by the roadside, practising ethical veganism, or even chanting at a football match, that exhibit fewer religious features. Once again, the black-and-white division between 'religion' and 'not-religion' has been replaced with a more complex landscape, some parts of which clearly demonstrate more religious 'features' than others.

In summary, it is by building on the example of Harrison's historically shifting *territories* of science and religion, and adopting and adapting this metaphor in the light of Wittgenstein's ideas of 'language as usage' and 'family resemblance', that we arrive at the idea of the *landscapes* of science and religion, and in so doing start to circumvent the essentialism that afflicts and encumbers on-going debates on this topic.

The foundation and structure of this book

This book hopes, on the basis outlined above, to bring a little bit of clarity to the evergreen debates around science and religion. It explores the relationship between the two, on the basis of how those terms are used and understood by a wide range of people today, and what features emerged from the ensuing process of disambiguation. In doing so, it allows us to approach afresh the overall question concerning science and religion: what exactly are we disagreeing about?

In this regard, the essay is in the same vein as David Bentley Hart's book on *The Experience of God*. This was prompted by his conviction that the concept of God, over which there had been so much recent noise, had remained obscure. 'The more scrutiny one accords these debates, the more evident it becomes that often the contending parties are not even talking about the same thing'. The disagreements, such as they are, are grounded in fundamental misunderstandings, and 'one cannot really have a disagreement without some prior agreement as to what the basic issue of contention is'.[68] Hart's basic issue of contention is 'God'; ours, less ambitiously, is 'science' and 'religion'. In Hart's terms, this book is an attempt to help us have a better (dis)agreement regarding science and religion.

It does this by drawing on a great deal of new data, generated by a major study into science and religion. Between 2019 and 2022, the authors conducted a research project into public and elite understandings of science and religion, which was funded by Templeton Religion Trust. The research comprised two elements.

The first was a qualitative study based on over a hundred in-depth interviews with the leading thinkers and public communicators in the fields of science and of religion, and in the field of science *and* religion, in the UK. This involved interviewing a wide range of experts—including philosophers and sociologists of science and of religion; people who work in astronomy, astrophysics, and physics; evolution, biology, and genetics; mathematics, chemistry, and engineering; psychology, sociology, and anthropology; religious studies and theology, as well as those with a particular cross-disciplinary interest in science and religion. Full details (including a list of interviews and a demographic and attitudinal breakdown) are given in appendix 1, but it is worth noting here that the academics interviewed were almost all in senior posts (50 of 66 in academic employment were professors). In addition to these interviews, we conducted more with communicators and popularisers of science and of religion, although there was considerable overlap between the academics and the popularisers. Altogether, the 101 interviews generated just short of a million words of transcripts. Quotations from these interviews are peppered throughout the book, identified by a number (e.g. #33) which refers to the order in which people were interviewed rather than in which they are listed in the appendix. While many interviewees said they were happy to be on the record, some were not, and to avoid confusion any details that might reveal the identity of an interviewee have been removed from the text.

The second element of the research was a quantitative study of a nationally representative sample of over 5,000 UK adults, assessing their understanding and opinion of science, religion, and science and religion. This was conducted by YouGov between 5 and 13 June 2021, after an iterative process of questionnaire design. The study interviewed a total of 5,153 respondents from the YouGov Plc UK panel of 800,000+ individuals. The sample was weighted so as to be representative of the UK (16+) adult population according to age interlocked with gender, social grade, ethnicity, and region. Full sample details are given in the appendices. The questionnaire asked respondents 26 numbered questions (plus various demographics), with

20   THE LANDSCAPES OF SCIENCE AND RELIGION

235 variables in total, which explored people's knowledge, understanding, and opinion of science and of religion and then, in detail, about their perception of the relationship between the different various disaggregated dimensions derived from earlier qualitative interviews.[69]

It is from this veritable mountain of data that this book constructs its arguments. Following this introductory chapter, the book is structured in two parts: part I comprising five chapters, part II, four. Chapter 2 traces the various attempts to define science in the UK and US, since its formation as a recognisable discipline distinct from natural philosophy in the nineteenth century. Drawing on philosophical sources (primarily the work of Robert Merton, Karl Popper, Larry Laudan, and Massimo Pigliucci) and legal ones (including the Overton Judgement, *Edwards v. Aguillard*, *Kitzmiller v. Dover Area School District*, and the British Science Council's 2009 definition of science), it shows how a singular, 'necessary and sufficient' definition of science is not possible, and that, following Pigliucci, a Wittgensteinian 'family resemblance' approach is the most appropriate and coherent way to 'define' science.

The chapter highlights distinct philosophical, methodological, cultural, and social dimensions or 'features' of science, and this process of disambiguation is developed in chapter 3, which draws on the data of the qualitative interviews. The analysis in this chapter identifies six distinct 'features' of science, pertaining to its (1) method(s), (2) subject, (3) presuppositions, (4) objectives, (5) values, and (6) institutional structure. Between these and the features identified in the legal and philosophical definitions discussion in chapter 2, we are able to build up a broad, composite, and flexible 'family definition' of what science is.

Chapters 4 and 5 go through the same process for religion. Chapter 4 outlines a number of higher profile and influential attempts to define religion over the last hundred years or so, drawing on various legal, anthropological, philosophical, and sociological contributions. Once again, but even more obviously, a single necessary and sufficient definition is beyond reach. Chapter 5 builds on this and draws on the new research to identify five distinct 'features' of religion, pertaining to (1) transcendence, (2) belief, (3) commitment, (4) the formalisation of belief in theology and doctrine, and (5) the formalisation of commitment in rituals, laws, and offices.

Chapter 6 turns its attention to the other new data source, namely the quantitative study of UK adult opinion. Building on the work of John Evans, who has argued for the ethical salience of the science and religion debate in the US, the chapter shows how this is indeed a significant factor within

UK public opinion. Public conceptions of science in the UK are usually embedded in progressive narratives, tied up with science doing things like warning, healing, or predicting. By contrast, religion—or more specifically religions—are also encountered as embedded in newsworthy narratives, although often ones that are problematic or even threatening. As a result, public conceptions of science and religion are usually weighted, in different directions, by their perceived ethical features. That recognised, the quantitative research conducted for this book also shows that, ethical salience notwithstanding, UK public conceptions of science and of religion, and their relationship, also recognise the significance of their metaphysical and methodological/epistemological features. The general public may have a more acute sense of science and religion tension in ethical/social/public arenas, but that does not come entirely at the cost of other, more philosophical, features. A brief concluding section summarises the findings of part I.

Part II then looks at the key areas or features within these disambiguated 'family definitions' of science and religion that come into contact and perceived conflict. It identifies and focuses on four: metaphysics, methodology, anthropology, and public authority and reasoning. In each instance, drawing primarily on the expert interviews but also, where appropriate, public opinion data, the chapters set out exactly what it is that people disagree about, whether that is, for example, science's apparent commitment to (some form of) naturalism coming into conflict with religion's apparent commitment to (some form of) 'supernaturalism', or science's much-hyped method(s) of organised scepticism coming into conflict with religion's apparent dependence on institutionalised faith.

In each case, the nature and strength of the perceived disagreement is set out, before the counter arguments, also drawn from the data, are outlined. Time and again, it seems that what initially appear to be insurmountable differences between science and religion can be closed up, disagreements negotiated, and conflicts defused. A great deal of the time, when people are disagreeing about science and religion, they do not need to be. That said, each chapter also recognises that, even after such diplomatic negotiations, there do remain points of tension, or 'outstanding questions' as the conclusion of each chapter puts it. These merit the attention of anyone seriously interested in on-going debates around science and religion. We leave these to the readers as points for further discussion.

There is no view from nowhere and it would be disingenuous to pretend that either author is neutral within this debate. Both of us believe that the debate between science and religion is interesting, worthwhile, and

22 THE LANDSCAPES OF SCIENCE AND RELIGION

important. Both are convinced that it can be fruitful, and both believe that the well-worn caricature, in which the two are implacably in conflict with each other, is groundless. One of the authors has recently written a history of science and religion that argues against the narrative of historical conflict, and we believe that just as the history of science and religion is not one of relentless antagonism, neither is its present.

However, we both also recognise that there are various legitimate points of contention here, areas where (*contra* Stephen Jay Gould) the two entities do overlap and (*contra* innumerable religious apologists) they do come into conflict. Just as the history of science and religion has been marked by moments of serious tension, so is its present. Readers, or at least those that follow us this far in our argument, will differ over which points of tension are more or less genuine, which are more or less serious, and which are more or less (in)surmountable. Accordingly, they will also disagree over what that means for science and religion. To attempt those questions would be to take us beyond the scope of this book. Our chosen scope is sufficient in itself. In a debate not always renowned for careful, precise, and respectful disagreement, we hope that our survey of the complex, subtle, varied landscapes of science and religion might help us understand better not only what each of these two human activities actually are, but also, therefore, what people are actually disagreeing about when they disagree about them.

Notes

1. Whitehouse, Harvey, François, Pieter, Savage, Patrick E., et al., 'Retraction Note: Complex societies precede moralizing gods throughout world history', *Nature* (7 July 2021). Reported in *The Daily Telegraph* as 'Small societies developed the "moralising God" concept to help them grow, claim academics', by Gabriella Swerling, 8 July 2021.
2. Falk, Dan, 'Learning to live in Steven Weinberg's pointless universe', *Scientific American* (27 July 2021).
3. Brigham and Women's Hospital, 'Brain circuit for spirituality?', *Science Daily* (1 July 2021).
4. Stamouli, Nektaria, 'Science vs. religion as Greek priests lead the anti-vax movement', *Politico* (20 July 2021).
5. Google Ngram Viewer.
6. Coyne, Jerry A., *Faith Versus Fact: Why Science and Religion Are Incompatible* (New York, 2015).
7. Gingras, Yves, *Science and Religion: An Impossible Dialogue* (Cambridge, 2017).
8. Dawkins, Richard C., *The God Delusion* (London, 2006).
9. Stenger, Victor, *God: The Failed Hypothesis: How Science Shows That God Does Not Exist* (Amherst, NY, 2007).
10. In many books, most recently *Natural Philosophy: On Retrieving a Lost Disciplinary Imaginary* (Oxford, 2022).
11. Collins, Francis S., *The Language of God: A Scientist Presents Evidence for Belief* (New York, 2006).

INTRODUCTION 23

12. Again, in many books, but for an overview see Polkinghorne, J.C., *The Polkinghorne Reader: Science, Faith, and the Search for Meaning* (London, 2011).
13. McLeish, Tom, *Faith and Wisdom in Science* (Oxford, 2014).
14. In many books, for example Deane-Drummond, Celia, *Christ and Evolution: Wonder and Wisdom* (London, 2009).
15. See Gould, S.J., *Rocks of Ages: Science and Religion in the Fullness of Life* (New York, 1999). It is worth noting that Gould modified the independence of NOMA. See Gould, S.J., *The Hedgehog, the Fox, and the Magister's Pox: Mending the Gap between Science and the Humanities* (London, 2003) and also McGrath, Alister. 'A consilience of equal regard: Stephen Jay Gould on the relation of science and religion', *Zygon* (2021), 56(3), 547–565.
16. McGilchrist, Iain, *The Master and His Emissary: The Divided Brain and the Making of the Western World* (New Haven and London, 2009).
17. McGilchrist, Iain, *The Matter with Things: Our Brains, Our Delusions, and the Unmaking of the World* (London, 2021).
18. Barbour, Ian G., *When Science Meets Religion: Enemies, Strangers, or Partners?* (San Francisco, 2000).
19. Haught, J.F., *Science and Religion: From Conflict to Conversation* (New York, 1995).
20. McFarland, Ian, 'Conflict and compatibility: some thoughts on the relationship between science and religion', *Modern Theology* (2003) 19, 181–202.
21. Stenmark, Mikael, *How to Relate Science and Religion: A Multidimensional Model* (Grand Rapids, MI, 2004).
22. Four of his models assume that a war is taking place: (1) scientism; (2) scientific imperialism; (3) theological authoritarianism; and (4) the evolution controversy. Six of the models assume a truce or some kind of partnership: (5) the Two Books; (6) the Two Languages; (7) ethical alliance; (8) dialogue accompanied by creative mutual interaction; (9) naturalism; and (10) theology of nature. See Peters, Ted, 'Science and religion: ten models of war, truce, and partnership', *Theology and Science* (2018) 16, 11–53.
23. Evans, John H. and Evans, Michael S. 'Religion and science: beyond the epistemological conflict narrative', *Annual Review of Sociology* (2008), 34, 87–105.
24. See Principe, Lawrence M., 'The Warfare Thesis' and Brooke, John Hedley 'Historians', in Hardin, Jeff, Numbers, Ronald, and Binzley, Ronald A. (eds.), *The Warfare between Science and Religion: The Idea that Wouldn't Die* (Baltimore, 2018).
25. The list of relevant texts is enormous and growing but the key ones include the following: Lindberg, David C. and Numbers, Ronald L. (eds.), *God and Nature: Historical Essays on the Encounter between Christianity and Science* (Berkeley and London, 1986); Brooke, John Hedley, *Science and Religion: Some Historical Perspectives* (Cambridge, 1991); Harrison, Peter, *The Fall of Man and the Foundations of Science* (Cambridge, 2007); Larson, Edward J., *Summer for the Gods: The Scopes Trial and America's Continuing Debate over Science and Religion* (New York, 2008); Dixon, Thomas, Cantor, Geoffrey, and Pumfrey, Stephen (eds.), *Science and Religion: New Historical Perspectives* (Cambridge, 2010); Brooke, John Hedley and Numbers, Ronald L. (eds.), *Science and Religion Around the World* (Oxford, 2011); Harrison, Peter, *The Territories of Science and Religion* (Chicago, 2015); McGrath, Alister E., *The Territories of Human Reason: Science and Theology in an Age of Multiple Rationalities* (Oxford, 2019); Lightman, Bernard, (ed.), *Rethinking History, Science, and Religion: An Exploration of Conflict and the Complexity Principle* (Pittsburgh, 2019).
26. Ungureanu, James C., *Science, Religion, and the Protestant Tradition: Retracing the Origins of Conflict* (Pittsburgh, 2019).
27. According to public opinion research conducted in June 2021, 58% of UK adults considered science and religion to be, on balance, incompatible (compared with 29% who thought them net compatible) and 23% considered the relationship between science and religion to be *strongly* incompatible, compared with just 7% who judged it strongly compatible. Similarly, when the Pew Forum surveyed 18 countries in Central and Eastern Europe in 2015–16, it found that in nine of them, more than half of respondents said there was a conflict between science and religion (see Pew Research Center,

24 THE LANDSCAPES OF SCIENCE AND RELIGION

https://www.pewresearch.org/religion/2017/05/10/science-and-religion/.) In reality, the *actual* public perception of the relationship between science and religion is rather more complex than these bald figures suggest—a fact that has been firmly established in the US, where there has been sustained research into public opinion about science and religion, in particular by Elaine Howard Ecklund and her team. See Ecklund, Elaine Howard, *Science vs. Religion: What Scientists Really Think* (Oxford, 2010) and Ecklund, Elaine Howard and Scheitle, Christopher P., *Religion vs. Science: What Religious People Really Think* (Oxford, 2017).

28. Hardin et al., *Warfare*, p. 1.
29. Lindberg and Numbers, *God and Nature*, p. 20 ('we must endeavour to understand the complexity and subtlety of their interaction').
30. Dixon et al., *Science and Religion*, p. 263.
31. Lightman et al., *Rethinking History*, p. 235.
32. Lightman et al., *Rethinking History*.
33. Harrison, Peter, 'Constructing the boundaries', *The Journal of Religion* (2006), 86(1), 81–106.
34. Harrison, *Territories*, pp. 11–16.
35. Harrison, *Territories*, pp. 7–11.
36. Midgley, Mary, *Science and Poetry* (London, 2001), pp. 81–87.
37. Quoted in Hunter, James Davison and Nedelisky, Paul, *Science and the Good: The Tragic Quest for the Foundations of Morality* (New Haven, 2018), p. 98.
38. Gould, *Rocks of Ages*, p. 6.
39. Glennan, Stuart, 'Whose science and whose religion? Reflections on the relations between scientific and religious worldviews', *Science & Education* (2009), 18, 797–812.
40. McGrath, *Territories of Human Reason*, p. 6.
41. Baker, Joseph O. 'Public perceptions of incompatibility between "science and religion"', *Public Understanding of Science* (2012), 21, 340–353.
42. Jong, Jonathan, 'On (not) defining (non)religion', *Science, Religion and Culture* (2015), 2(3), 15–24.
43. Laborde, Cécile, *Liberalism's Religion* (Cambridge, MA, 2017), pp. 18, 5; emphases added.
44. Laborde, *Liberalism's Religion*, p. 3.
45. See, for example, Bardon, Aurélia and Howard, Jeffrey William (eds.), *Liberalism's Religion: Cécile Laborde and Her Critics* (London, 2020).
46. Plantinga, Alvin, *Where the Conflict Really Lies: Science, Religion, and Naturalism* (Oxford, 2011). See also his debate with Dennett in Plantinga, Alvin, and Dennett, Daniel C., *Science and Religion: Are They Compatible?* (New York, 2011).
47. Plantinga, *Where the Conflict Really Lies*, p. ix.
48. Evans, John, *Morals Not Knowledge: Recasting the Contemporary U.S. Conflict between Religion and Science* (California, 2018).
49. Asma, Stephen, *Why We Need Religion* (New York, 2018).
50. Elson-Baker, Fern and Lightman, Bernard, *Identity in a Secular Age: Science, Religion and Public Perceptions* (Pittsburgh, 2020), pp. 37, 65.
51. Stenmark, *How to Relate Science and Religion*, pp. xvi–xx.
52. Berger, P.L., *The Sacred Canopy* (Garden City, NY, 1967).
53. The idea that religion is a complex, socially constructed entity in need of disaggregation is familiar within the literature, even if there is no consensus on its constituent elements. The idea that science is similarly complex or socially constructed is less so, albeit this is familiar to philosophers and sociologists of science. Nevertheless, as we shall see, science can be disaggregated every bit as thoroughly and fruitfully as religion.
54. Jong, 'On (not) defining (non)religion', p. 17.
55. Ritual, narrative, experiential, social, ethical, doctrinal, and material.
56. As we shall see, a similar problem can be posed with regard to science. Were we, for example, to agree on the Science Council's definition of science, would an activity be legitimately classified as science if it incorporated observation and used evidence and critical analysis, but not measurement, experiment, and repetition? Conversely, if an activity

integrated all the requisite dimensions from the definition, should we still classify it as science, even if it is, say, demonstrably the work of historical research or journalism or police detection?

57. Wittgenstein, Ludwig, *Philosophical Investigations* (Oxford, 1999[1953]), #309. Given Wittgenstein's famously gnomic utterance, this, on the surface of it, seems like a rather hopeful venture. See also Labron, Tim, *Science and Religion in Wittgenstein's Fly Bottle* (New York, 2017).

58. Harrison, *Territories*, p. 184.

59. McGrath, *Territories of Human Reason*, p. 5.

60. Laborde, *Liberalism's Religion*, p. 30.

61. Wittgenstein, *Philosophical Investigations*, #1.

62. Wittgenstein, *Philosophical Investigations*, #66.

63. 'Giving orders, and obeying them—Describing the appearance of an object, or giving its measurements—Constructing an object from a description (a drawing)—Reporting an event—Speculating about an event—Forming and testing a hypothesis—Presenting the results of an experiment in tables and diagrams—Making up a story; and reading it—Play-acting—Singing catches—Guessing riddles—Making a joke; telling it—Solving a problem in practical arithmetic—Translating from one language into another—Asking, thanking, cursing, greeting, praying'.

64. The example is deliberately chosen because of the employment ruling in the case of Jordi Casamitjana in January 2020. Employment Tribunals, Case Number: 3,331,129/2018, *Mr J Casamitjana Costa* v. *The League Against Cruel Sports*, 2–3 January 2020. The disaggregated elements cited here derive from Mr Justice Burton's summary of the Strasbourg definition of 'philosophical belief' in *Grainger PLC* v. *Nicholson* [2009] UKEAT 0219/09/ZT (3 November 2009), para. 24.

65. Emphases added.

66. Wittgenstein, *Philosophical Investigations*, #68, #70.

67. Wittgenstein, *Philosophical Investigations*, #69.

68. Hart, David Bentley, *The Experience of God*: *Being, Consciousness, Bliss* (London, 2013), p. 2.

69. All within and tailored to the capacity of a nationally represented quantitative sample, of course. We didn't ask many questions about epistemology, at least not in so many words.

PART I

SURVEYING THE LANDSCAPES OF SCIENCE AND RELIGION:

Defining and Disaggregating Terms

2
Defining Science

The need to define science

The British have never felt a particularly pressing need formally to define science. In 1843, Parliament passed the Scientific Societies Act, which exempted from the payment of rates all 'societies established exclusively for purposes of science, literature or the fine arts'.[1] The Act has remained on the statute book ever since, but in spite of its age, there hasn't been a single case in a British law that has sought to define 'science' for the purposes of the Act.[2]

Its redundancy has not been complete. In 1999, the High Court in Ireland heard a case between the Transport Museum Society of Ireland and Registrar of Friendly Societies concerning a lower court's refusal to grant the former the benefit of a certificate under the 1843 Act. The court was required to determine whether the Transport Museum Society was indeed engaged in the practice of science for the purposes of the Act. In doing so, it took a 'broad interpretation' of what science was. Referring to an earlier judgement from Lord Denning, the Court claimed that science included 'not only pure or abstract science, such as pure mathematics' but also 'applied sciences, such as electricity or engineering', and that 'the purposes of science' meant not only 'the advancement of knowledge by research work' but also 'the dissemination of it by lectures or teaching or writing'. In effect, taking its cue from the Act's full phrase 'science, literature or the fine arts', the court ruled that science should be interpreted in line with its historic and etymological roots—of 'knowledge' in the widest sense—and not in the narrow 'academic sense as covering mathematics, chemistry, physics, and biology' alone.[3] Science, in other words, was a capacious category.

The situation in America has been somewhat different. On account of the prevalence of what is called 'creation science', the need formally to define science in US law has been more urgent and less instinctively generous. When, in 1981, the Governor of Arkansas signed into law an Act that mandated that public schools in the state 'shall give balanced treatment to creation-science and to evolution-science', the inevitable ensuing trial, resulting in

30 THE LANDSCAPES OF SCIENCE AND RELIGION

the so-called Overton Judgement, paid particular attention to the question of what constituted science. Unlike in Ireland, where the rateable status of the Transport Museum in Howth was at stake, in Arkansas it was the somewhat more vexed issue of children's education.

All the evidence suggests that the general direction of travel, legally and culturally, in the US, UK, and Europe, is towards Overton rather than Howth—namely the need to define science and to do so with greater precision and constraint. In the UK, the 2006 Charities Act for the first time included 'the advancement of science' as a charitable activity. Scientific organisations were now in the same position as many religious and cultural ones in being eligible for charitable status, with its benefits of tax relief. The Charities Act itself offered little more in terms of definition than did the 1843 Act, stating only that 'science includes scientific research and charities connected with various learned societies and institutions'.[4] This, plus the separation of 'the advancement of science' from accompanying activities, such as 'literature or the fine arts' in the 1843 Act, means there is a good chance that the meaning of science will one day need to be clarified in the UK courts.

Recent years have also seen a significant politicisation of science that has necessitated an increased degree of self-reflection.[5] In her Tanner Lectures on Human Values, subsequently published as *Why Trust Science?*, Naomi Oreskes reflected on the growing public distrust in science in the Western world and in the US in particular.[6] This, she argued, was driven by a variety of factors: tobacco companies sowing disinformation about the medical effects of their products; free-market think tanks and the oil industry casting doubt on the reality and extent of anthropogenic climate change; public mistrust in government being transferred to any politically mandated health measures like mass vaccination; and fundamentalist and 'creation science' groups orchestrating campaigns against Darwinian evolution and its teaching in schools.

The response has been equally varied—from eager political acquiescence to angry cultural bellicosity—but in its most sensitive and intelligent forms, of which Oreskes' book is an example, it has provoked a careful and judicious exercise in self-reflection. '*Should* we trust science? If so, on what grounds and to what extent?'[7] The process of answering these questions invites philosophical reflection on the basis and reliability of the scientific method (or, more accurately, scientific methods) *and* sociological reflection on the actual practice of scientists. Between them, these approaches amount to a sustained attempt to answer the wider question 'what is science?'

The 'crisis' in public trust in science—the word merits scare quotes as scientists are still trusted rather more than most other professionals—identified by Oreskes is primarily of external making—business, think tanks, anti-vaxxers, and fundamentalists casting the doubt. There have been other catalysts to define science that are arguably more of science's own making. Indeed, some are the result of science's *success*.

The success of natural science has generated an authoritative methodological paradigm that is now eagerly imitated by many other comparable disciplines, which often claim a similar ability to predict (and manipulate) reality. However, as Jason Blakely argued in his book on the rise of social science, this imitation has proven conspicuously fallible. The subject of social science, the human world, is not as amenable to the same form of 'objective' analysis as is the natural world. 'Human social and political behaviour does not fit under the conceptual logic of the natural sciences because it is not law-abiding or mechanistic in nature.'[8] The result is a number of well-publicised social and political scientific errors—such as repeated psephological failures to predict election results or the almost total failure of economic science to predict the biggest crash in 80 years—that have cast doubt not only on the reliability of social science itself, but also on the scientific method(s) on which it purports to rest.

The same pattern of flattery and imitation was evident in the famous 'hoax' in which Alan Sokal and others sent an academic paper, entitled 'Transgressing the boundaries: towards a transformative hermeneutics of quantum gravity', to *Social Text*, a journal of postmodern cultural studies.[9] Sokal, a professor of physics, marinated the paper in dense scientific language, and argued that quantum gravity was merely a social and linguistic construct. The argument was, however, pure nonsense, intended to expose the vacuity of the postmodern turn in humanities. The affair obviously reflected worst on the journal that accepted Sokal's paper and the intellectual discipline it represented. However, the entire affair was driven precisely by (what Sokal called in another book) 'postmodern intellectuals' abuse of science', which, although less effective than social science's appropriation of natural science's authority, still had the effect of muddying the waters around what science is and what it says.[10]

In a similar vein, but at a still more egregious level, the success of science has catalysed the growth of straightforwardly pseudoscientific disciplines. Pseudoscience is not new; an English word dating back to the 1840s, coming from the Latin *pseudoscientia*, was already in use in the first half of the seventeenth century. However, as Michael Gordin has shown, it is in

32 THE LANDSCAPES OF SCIENCE AND RELIGION

particularly good health today, sticking to science like its own shadow and growing in size as the stature and authority of science grows. The persistence and/or emergence of astrology, UFOlogy, parapsychology, 'scientific' vaccine-rejection, and the like have provided a powerful stimulus to the age-old demarcation problem, which seeks to distinguish science from non- and pseudo-science.[11]

There have been causes for concern from within science too, with internal practices and structures threatening to erode or distort science's fundamental capacity to challenge and revise received knowledge. In 2004, Henry Bauer, a professor of chemistry at Virginia Polytechnic Institute and State University, published a paper on the stifling effect that institutionalisation and commercialisation were having on original research.[12] 'Science is now altogether different from the traditional disinterested search, by self-motivated individuals, to understand the world', he wrote. 'What national and international organizations publicly proclaim as scientific information is not safeguarded by the traditional process of peer review', with the result that society now needs new ways of guaranteeing the trustworthiness of public information about science.

A further catalyst for self-reflection and delineation has come from what is widely known as the 'replication crisis'.[13] On the basis that replicability has been a cornerstone of science since the time when it was natural philosophy, an increasing number of studies have shown how, in certain disciplines, the proportion of replicable papers is worryingly low. Stuart Ritchie charts this in his book *Science Fictions*, reporting how one study published in *Nature* in 2015 reported that only 39% of studies in psychology, of 100 attempted, were judged to have been successfully replicated. Another, of a smaller sample of social science papers that had been published in *Nature* and *Science*, which was published three years later, found the replication rate was 62%. Almost all studies, even when successfully replicated, found that the original studies had exaggerated the size of their effects.[14] Psychology and social science were, it seemed, particularly badly affected, but there was evidence of similar problems in economics, neuroscience, evolutionary biology, organic chemistry, and medical science. The problem has helped catalyse a new discipline of meta-science, scientific research about scientific research, and further reflection of what practices constitute legitimate science.

There are some signs of even more serious problems underlying all this. In his book *Fraud in the Lab*, the investigative science journalist Nicolas Chevassus-au-Louis charted the apparent growing levels of outright deception within scientific practice.[15] The rate of retraction of scientific papers has risen dramatically over the last few decades, increasing eleven-fold between 2001 and 2010 for instance. This is alarming but needs to be put into perspective, as retraction rates remain very low, at about 0.02%. Moreover, not all retractions indicate nefarious processes (by one measure, only 26% of retracted articles are down to fabrication or falsification[16]) and, moreover, retraction itself can legitimately be interpreted not as a problem but as an indication of science's essential self-correcting method, a sign of scientific health rather than pathology. Nevertheless, all that recognised, the rate of increase in retraction; the correlation between a journal's impact factor and its retraction rate; numerous sociological studies into laboratory practices and reported levels of malpractice; together with the absence of any reason for a comparable rise in 'good-faith' retractions, all suggest that there is a real problem here for science.

As with pseudoscience or replication, this is not a new problem,[17] nor is it evenly spread—mathematics and physics proving to be largely immune, medicine rather less so. But it is a problem, demanding careful response. Thus, the US Office of Science and Technology Policy defined breaches in scientific integrity in 2000,[18] while the Organisation for Economic Co-operation and Development (OECD) set out best practices 'for ensuring scientific integrity and preventing misconduct' in 2009.[19] Defining scientific misconduct, however, naturally involves having some idea of what scientific conduct and, more broadly, science itself is, or at least what it should be. Again, we circle round to the central issue, namely the need to establish a clear idea of what science is, both in theory and in practice.

The attempts to define science

This confluence of legal and cultural challenges has precipitated several significant and formal attempts to define science, but it is worth noting that they do not come *ex nihilo*. Attempts to define science arguably date back to Aristotle's efforts in *Posterior Analytics*. Indeed, in as far as science historically meant simply 'knowledge', the attempt to define it, and in particular to distinguish warranted and trustworthy knowledge from its alternative, has been a centrepiece of philosophy since its inception.

In the early nineteenth century, the mathematician Charles Babbage published his *Reflections on the Decline of Science in England*, in which he fingered hoaxing, forging data, trimming data, and cooking data as the causes of science's alleged decline.[20] His (much disputed) criticisms catalysed the

34 THE LANDSCAPES OF SCIENCE AND RELIGION

foundation of the British Association for the Advancement of Science and, indirectly, informed the work of the philosopher and theologian William Whewell, who helped establish, professionalise, and systematise 'science' as a distinct, autonomous, and respectable discipline, partly through his landmark *Philosophy of the Inductive Sciences*, and partly through coining the word 'scientist' (as well as numerous other scientific terms).[21] In short, questions about what (modern, professional) science was, what it wasn't, and how it should be identified and legitimised attended its very birth.

Such questions matured in the early years of the twentieth century, as disciplines like Freudian psychoanalysis and Marxist history confidently appropriated the label of science, prompting philosophers, most famously Karl Popper, to try to establish what separated science from pseudoscience, in what became known as the 'demarcation problem'.[22] A few years later, around a century after Whewell began to systematise science, the sociologist Robert Merton set out four scientific values to which science was expected to aspire. His essay, originally published as 'Science and Technology in a Democratic Order', claimed that science was (or should) be characterised by a commitment to universalism (science is science irrespective of who participates), disinterestedness (science is not motivated by ideology and material reward or social recognition), communism (which is perhaps better rendered as communality—science shares knowledge), and organised scepticism (science questions received wisdom but in a structured way).[23]

These values would prove influential and come to be known as Mertonian norms, but they were not, obviously, the last word on the subject of 'the normative structure of science', the title of Merton's essay when it was republished. As the cultural authority of science and technology rose in the post-war period, ever more disciplines attempted to assume that authority by appropriating its language and methods. The question of what science was became socially and politically urgent.

The resurgence of anti-evolutionary views in America from the 1960s was accompanied by a lexical shift, in which the whole movement, its institutions, publications, and arguments, appropriated the language and framing of science. Biblical literalism was now presented through the endeavours of the Institute for Creation Research, the Creation Science Research Center, and the Creation Research Society.[24] Their success in animating public opinion resulted in attempts to put 'creation science' on the curricula, ideally on a parity with the teaching of evolution.

DEFINING SCIENCE   35

The resulting case in Arkansas stated emphatically that creation science was not science, for which it offered both a sociological and a philosophical justification. Regarding the former, the court ruled that a 'descriptive definition' of science is what is 'accepted by the scientific community' and 'what scientists do'. Regarding the latter, it stated, more fully, that the 'essential characteristics' of science are:

1. it is guided by natural law;
2. it has to be explanatory by reference to natural law;
3. it is testable against the empirical world;
4. its conclusions are tentative, i.e. are not necessarily the final word; and
5. it is falsifiable.[25]

This judgement made a clear (although far from final) distinction between 'evolution science' and its alleged alternatives, and in so doing provided a robust and confident definition of science, and a demarcation between that which could legitimately be categorised as science and that which could not.

This, however, was a legal position, not a philosophical one, and the following year, the American philosopher Larry Laudan published an important essay entitled 'The Demise of the Demarcation Problem'. This argued that demarcation was a doomed enterprise and there was no hope of generating a definition of science that was simultaneously necessary and sufficient.[26]

Laudan traced attempts to define science from the Enlightenment conception that science was uniquely apodictic, essentially coterminous with infallible knowledge, through a fallibilistic perspective that shifted emphasis onto allegedly distinctive methodological considerations, and into positivist and Popperian attempts in the mid-twentieth century, which adopted 'syntactic and semantic strategies' for adjudicating what was meaningful, truthful, and scientific. His judgement, particularly on this 'new demarcationist tradition', was withering. 'The evident epistemic heterogeneity of the activities and beliefs customarily regarded as scientific should alert us to the probable futility of seeking an epistemic version of a demarcation criterion.'[27] As a result, he concluded, whereas the question of what made a belief well-founded or heuristically fertile was both interesting and tractable, the parallel question, into which it was often elided, 'what makes a belief scientific?' was neither. The '"scientific" status of [such] claims is altogether irrelevant'.[28]

36   THE LANDSCAPES OF SCIENCE AND RELIGION

Laudan's arguments were to prove influential, and academic interest in the demarcation problem waned over the next 30 years. Popular interest, however, particularly in America, did not. In 1987, the Supreme Court heard a case, *Edwards v. Aguillard*, in which the constitutionality of a Louisiana state law requiring the teaching of 'creation science' alongside evolution was decided. As part of the case, 72 Nobel laureates, 17 State Academies of Science, and 7 other scientific organisations made a joint written submission to the court in support of the appellants.[29] This, among other things, included a number of paragraphs that outlined a definition of science that was less systematic and concise but more detailed than that of the Overton ruling.

According to their submission, science was based on an approach to acquiring knowledge ('the Scientific method') which involves 'the rigorous, methodical testing of principles'. It prizes consistency—'to be a legitimate scientific "hypothesis", an explanatory principle must be consistent with prior and present observations'—but also openness—'[a hypothesis] must remain subject to continued testing against future observations'. Its currency is facts, hypotheses, and theories. It involves 'systematically collecting and recording data about the physical world', which it then categorises and studies 'in an effort to infer the principles of nature that best explain the observed phenomena'. Its hypotheses about this must be tested ('an explanatory principle that by its nature cannot be tested is outside the realm of science') and those that 'accumulate substantial observational or experimental support ... become known as scientific "theories"'. Those theories that are able consistently to predict new phenomena that are subsequently observed are treated as 'especially "reliable" theory', although openness remains important even here and even the most reliable 'is forever subject to reexamination and ... may ultimately be rejected'.

Importantly, science is naturalistic. The facts with which it deals are the properties of natural phenomena, and it is 'devoted to formulating and testing naturalistic explanations for natural phenomena'. That means it is 'limited', 'not equipped to evaluate supernatural explanations' or passing judgement on their truth or falsity—an admission that can be read in rather different ways. In addition to all this, science is cumulative, achieving 'an ever increasing body of observations that give information about underlying "facts"', and it is distinguished by a distinct vocabulary, albeit 'individual scientists are not always careful in their use of that vocabulary'.

The judge in *Edwards v. Aguillard* found for the appellants, clearly recognising, as did the Overton ruling, that creation science did not meet these standards for science. Nevertheless, in spite of the fact that this was, unlike

Overton, a Supreme Court hearing, it was not the end of the matter, as 'creation science' morphed into Intelligent Design (ID), which returned to the courts nearly 20 years later.

In 2005, a District Court for the Middle District of Pennsylvania heard a case in which the status of ID as a scientific theory and legitimate alternative to evolution was debated.[30] Like *Edwards v. Aguillard*, this case did not result in as clear or definitive a statement on the nature of science as did the Overton ruling, but it did elicit, from the expert witnesses, a number of key characteristics that pointed in that direction—albeit characteristics that were shaped by the particular concerns of the case.

Science, the court heard, demands a 'rigorous attachment to "natural" explanations'. It is 'ruled by methodological naturalism'. It rejects 'the appeal to authority, and by extension, revelation' in favour of empirical evidence. Its subject is 'limited to empirical, observable and ultimately testable data'. It seeks explanations that are 'based upon what we can observe, test, replicate, and verify'. Those explanations are restricted to what 'can be inferred from the confirmable data', and the results, obtained through observations and experiments, 'can be substantiated by other scientists'. Testability, 'rather than any ecclesiastical authority or philosophical coherence', is the basis of its authority. It deliberately omits 'theological or "ultimate" explanations' and 'does not consider issues of "meaning" and "purpose" in the world'. On this basis, ID failed to qualify as a science, and the court found for the plaintiffs.

Science enjoyed another attempt at definition, albeit for happier reasons, the following year in the UK, as the 2006 Charities Act was passed, which prompted the Science Council to clarify what the word 'science' meant. Following a year-long consultation, the Council proposed a definition of science as 'the pursuit and application of knowledge and understanding of the natural and social world following a systematic methodology based on evidence'. Given the pivotal role of the word 'methodology' in the definition, the Council went on to define that 'scientific methodology' included:

1. objective observation: measurement and data (possibly although not necessarily using mathematics as a tool);
2. evidence;
3. experiment and/or observation as benchmarks for testing hypotheses;
4. induction: reasoning to establish general rules or conclusions drawn from facts or examples;
5. repetition;

38   THE LANDSCAPES OF SCIENCE AND RELIGION

6. critical analysis;
7. verification and testing: critical exposure to scrutiny, peer review, and assessment.[31]

This definition was published in 2009, around which time the philosophical deep-freeze into which the 'demarcation problem' had been placed after Laudan's essay began to thaw. In the first instance, this was on account of a collection of essays edited by Massimo Pigliucci and Maarten Boudry.[32]

Pigliucci and Boudry's volume was motivated by the aforementioned rise of pseudoscience—creationism, alternative medicine, conspiracy theories about AIDS, climate change denialism, cults and sects such as Scientology, psychoanalysis, and so on—to argue that there was not only a scholarly but also increasingly an ethical and social duty to be able to define, refute, and exclude pseudoscience. That being so, the volume turned the tables on the demarcation question by asking how pseudoscience, rather than science, might be defined, and addressed Laudan's objections by arguing that he had set the bar unrealistically high, thereby making his death pronouncement for the problem 'premature and misguided'.[33] Not being able to locate necessary and sufficient conditions for science should not be a barrier to tracing its approximate boundaries with pseudo (and non-)science.

Pigliucci's opening chapter homed in on a Wittgensteinian solution (or, more modestly, approach) to the problem. Drawing on a similar approach about 'the Disunity of Science' outlined by John Dupré 20 years earlier,[34] Pigliucci advocated what he termed 'the possibility of understanding science as a Wittgenstein-type cluster concept'.[35] Rather than demanding necessary and sufficient conditions, as had Laudan, science was better understood in terms of 'fuzzy [logic] and gradual distinctions'.[36] Fuzzy logic was developed as a way of capturing degrees of membership or of truth, as presented by problems such as the well-known Sorites paradox (at what point does a heap of sand stop being a heap if grains are removed one at a time?) By this reckoning, science is 'a matter of degree rather than kind'.[37]

The principle of Wittgenstein family resemblances added various dimensions to this fuzzy logic, and Pigliucci suggested two ('at a very minimum' as a way of 'start[ing] somewhere'): empirical support and theoretical understanding (i.e. internal coherence and knowledge). This results in a two-dimensional sliding scale in which various disciplines could be classified as more or less scientific. By his own reckoning, some activities, like particle physics, climate science, and molecular biology, represent established science because they are both empirically robust and theoretically coherent. Some, like economics, psychology, and sociology, should be classified as 'soft' sciences, on account of the fact that although empirically well testified, they are also marked by 'epistemic heterogeneity'. Others, such as the Search for Extraterrestrial Intelligence, superstring physics, some evolutionary psychology, and scientific approaches to history, are best considered as proto- or quasi-scientific because, although high on theory, they lack empirical warrant. And finally, some, such as ID theory, astrology, and HIV denialism, are clearly pseudoscience, for having little or no theoretical or empirical weight. In this way, a 'family resemblance' approach offered a way forward in the task of defining science, without being broken on the unrealistically hard rocks of 'necessary and sufficient' criteria or being sucked into the despairing whirlpool of postmodern woolliness.

A 'family resemblance' approach to defining science

The Science Council's definition of science was well received, commended by, among others, the philosopher A.C. Grayling and by David Edgerton, then professor of the history of science and technology at Imperial College. Their comments, however, were telling, further underlining the need for the 'family resemblance' approach to defining science already proposed by Pigliucci. Grayling's commendation was specifically on the basis that the definition was sufficiently 'general' to capture science's 'very wide range of activities', while Edgerton pointed out that 'a definition of science needs to define the nature of the knowledge not the means of its creation only', and that, even with its latitude, the Council's definition 'would include historical research and indeed some journalism', and fails to 'demarcate something called science from the humanities'.[38] Such is the fate of all attempts to define science. Decisive demarcation is impossible. Fuzzy logic, degrees of membership and family resemblance is the best option available.

This Wittgensteinian approach is central to our approach, as we shall see in greater detail in following chapters, but the manner in which it works can be seen by drawing on five of the definitions of science we have cited in this chapter: Mertonian norms, the Overton judgement, *Edwards v. Aguillard*, *Kitzmiller v. Dover Area School District*, and the Science Council definition. Between them these definitions generate a large number of overlapping

40 THE LANDSCAPES OF SCIENCE AND RELIGION

criteria which capture the breadth of what science is. Moreover, they can profitably, if tentatively, be categorised into groups—different branches of the family, as it were—a process that affords a little clarity in the family definition of science.

Thus, we can say, firstly, that science is defined by certain philosophical axioms or presuppositions. It is naturalistic: 'devoted to formulating and testing naturalistic. explanations for natural phenomena' (*Edwards*) or 'ruled by methodological naturalism' (*Kitzmiller*). It is committed to the idea of lawfulness: 'guided by natural law' (Overton). And it is 'limited' (*Edwards*), focusing on the 'natural and social world' (Science Council) and intentionally omitting 'ultimate explanations', and ignoring (but not therefore disproving) 'issues of "meaning" and "purpose" in the world' (*Kitzmiller*).

Second, it is characterised by a strict methodological approach. This approach necessitated the 'rejection of authority' (*Kitzmiller*). It involves the formation of hypotheses and the collection of empirical evidence or data through observation and/or experimentation. These data involve measurement (Science Council) and possible mathematical calculation (Science Council), by means of which the hypotheses are assessed. The process is replicable and the results confirmable (*Kitzmiller*) or falsifiable (Overton) or verifiable (*Kitzmiller*). The method is inductive, intending 'to establish general rules or conclusions drawn from facts or examples' (Science Council); explanatory, that is 'infer[ing] the principles of nature that best explain the observed phenomena' (*Edwards*); and ideally predictive.

All five sources show that the methodological criterion is central to what science is, but however central it is, it is not sufficient, and there is, thirdly, what might be called an attitudinal or cultural dimension to science. Science is a process of 'organised scepticism' (Merton). It is attitudinally open, 'subject to continued testing against future observations' and 'forever subject to re-examination' (*Edwards*). All its conclusions are tentative (Overton). It is disinterested (Merton), systematic (*Edwards*), and prizes consistency (*Edwards*) in its approach.

Finally, science is a social activity—a fact that is often overlooked in popular depictions but noted in all except the Overton ruling above. It is a communal and collaborative activity (Merton). Its conclusions can, indeed must 'be substantiated by other scientists' (*Kitzmiller*). Its activities are open to 'critical exposure to scrutiny, peer review and assessment' (Science Council). And its achievements are 'cumulative' (*Edwards*).

These four dimensions—philosophical, methodological, cultural, and social—comprise a range of criteria against which the 'science-ness' of an activity may be measured—those activities satisfying more criteria being more warranted in their appropriation of the label science than those with fewer.

This is the approach we will adopt in the following chapter when we come to look at the findings from the expert interviews conducted as part of this project, where we will see these four criteria and their details confirmed and expanded. However, before we do so, there is one final point regarding the approach to defining science which demands attention.

The awkward reality of science

Over the last 40 years or so, Richard Dawkins has proved to be one of the most forceful and outspoken ambassadors for science, albeit one sometimes indifferent to the diplomatic protocols of ambassadorial work. In his advocacy he has frequently told a story in which, as an undergraduate, he witnessed a 'respected elder statesman' of Oxford's Zoology Department, who had for years 'passionately believed' that a microscopic feature of the interior of cells was not real, being publicly presented with evidence to the contrary by a visiting American cell biologist. 'At the end of the lecture, the old man strode to the front of the hall, shook the American by the hand and said—with passion—"My dear fellow, I wish to thank you. I have been wrong these fifteen years". 'The memory of the incident', Dawkins remarked in *The God Delusion*, 'still brings a lump to my throat'.[39]

Fellow travellers have made similar points. Jerry Coyne explained in *USA Today* how

> Science operates by using evidence and reason. Doubt is prized, authority rejected. No finding is deemed 'true'—a notion that's always provisional—unless it's repeated and verified by others. We scientists are always asking ourselves, 'How can I find out whether I'm wrong?'[40]

There is much in Coyne's definition—evidence, scepticism, provisionality, repeatability, verifiability, community—that is recognisable in the family features of science assembled above. However, it would be a mistake to end the story, or the definition, here. Quoting Coyne, the Dutch primatologist

42   THE LANDSCAPES OF SCIENCE AND RELIGION

Frans De Waal has commented, 'Oh, how I wish I had colleagues like Coyne! Having spent all my life among academics, I can tell you that hearing how wrong they are is about as high on their priority list as finding a cockroach in their coffee'.[41]

In contradiction to Dawkins' misty-eyed and Coyne's mythical depiction of science—the first based on a sample size of one; the second, somewhat ironically, citing no empirical evidence at all—there is much hard evidence that however much science *should* proceed along the lines and embody the criteria outlined above, all too often it does not. In his defence, Dawkins does go on to admit that 'in practice, not all scientists would [do what the elder statesmen did]. But all scientists pay lip service to it as an ideal'. It is a telling admission, not least the way it turns the idea of simply 'pay[ing] lip service to ... an ideal' from being a criticism, which it usually is, to a commendation.[42]

Historians and sociologists of science have repeatedly shown that the theoretical design of science sometimes bears only a faint resemblance to what actually happens. There are many examples of (established and professional) science being utterly interested (as opposed to disinterested); of conforming observations to theory rather than the other way round; of reaching and holding firmly to conclusions that were repeatedly and with great effort shown to be wrong; of genuflecting before 'respected elder statesman' rather than listening to the insights of junior colleagues; and so forth. And all this, it should be emphasised, is without venturing into the territory of outright fraud discussed by Babbage and Nicolas Chevassus-au-Louis.

This is not a new issue for science. As long ago as 1908, the physicist and historian and philosopher of science Pierre Duhem showed that the methods actually used by practising scientists bore little resemblance to the scientific method that was so foundational to the idea of science at the time.[43] More recently, the emergence of what is known as 'scientific pluralism' is based precisely on the empirically rooted recognition that 'scientific theories often do not reduce, there is not one universal scientific method, not only one fundamental scientific ontology, and successful science requires not only epistemic but also social diversity'.[44] In his exploration of the contemporary replication crisis, Stuart Ritchie emphasises how science is an inherently 'social' and 'human' activity, which, despite being premised on a system of effective checks and balance, has (in some areas) become a victim of complacency, rhetoric, and perverse incentivisation that has ended up amplifying rather than winnowing out natural foibles.[45]

This is a practical problem, with evidence that scientists don't all live up to the high standards of what science should be. But it may be more than that. It may be that the very criteria of what science should be are self-contradictory. For example, it is hard to see how science can simultaneously be cumulative *and* entirely open; new research must, by definition, assume the truth of that which has gone before it. In a similar vein, it is not clear how science can altogether eschew authority and yet at the same time educate young scientists. Every lecture in which a lecturer speaks to a class rather than has them replicate experiments is, by definition, an exercise of authority. Every project with a principal investigator (PI), or lab with a professor, or official institute with a governing body or advisory council is in some way recognising and exercising a form of authority, and is usually the stronger for it. The very recognition of science as an institution presupposes a body of authority. Authority and experiment, for all they are opposed in the theoretical definition of science, have, in reality, a more complex relationship. In short, the theory and practice of science are far from entirely coterminous, and if we are serious about wanting to understand what science is, we need to take serious account of the 'is' as well as the 'ought', integrating a descriptive element alongside the normative one in our definition.

Tellingly, many of the respondents we spoke to fully recognised this. The sheer number of expert interviewees who (usually voluntarily and often in considerable detail) compared the actual practice of science to its theory was striking. Time and again, scientists (but also sociologists and philosophers of science, and science communicators) stressed how the reality of science complexified the (already complex) familiar normative definition of science. This could be negative—many interviewees talked about the biases, prejudices, assumptions, and egotism that can distort scientific work—but it could also be positive, with a number pointing out that creativity, intuition, and even inspiration played an important role in the practice of science.

Either way, it was clear that to understand science satisfactorily demanded not only incorporating the normative philosophical, methodological, cultural, and social dimensions within a family resemblance definition, but also taking into account the descriptive reality of science as actually practised. Given how much of the science and religion debate is based on a comparison between a normative understanding of science and a descriptive understanding of religion, this is a vital point.

THE LANDSCAPES OF SCIENCE AND RELIGION

Conclusion: defining science

This chapter has outlined the various needs and attempts to define science over the last 50 years or so. In doing so, it has highlighted not only how that need is more acute today than ever before but also how the various attempts to do so gravitate towards the kind of Wittgensteinian 'family resemblance' approach outlined in the introductory chapter.

In an age of anti-vaxxers, alternative medicine, conspiracy theories about AIDS, postmodern excess, climate change denialism, Intelligent Design, replication crisis, hubristic social science, and other non- or pseudo-scientific activities, we need at least to attempt to delineate warranted from unwarranted truth claims, and that means returning to the allegedly deceased 'demarcation problem'. In much the same way, if we ever want to achieve any clarity in the debate between science and religion, and to understand what exactly people are dis/agreeing about when they dis/agree about science and religion, we need a similarly robust, accurate, and realistic understanding of what science (and, as we shall see, what religion) is.

Doing so, however, is far from straightforward, as the various philosophical and legal attempts to do so through the twentieth century show. Identifying watertight, necessary, or sufficient components to the definition is impossible. Science is too complex, diverse, heterogeneous, social-mediated, and fluid an activity to lend itself to a single, simple definition, or even a set of definitions. Rather, between them, these attempts naturally gravitate to the idea of a family resemblance, whereby a set of characteristics—normative and descriptive—enable us to determine whether something is more or less 'science', without compelling us to draw an absolute, impermeable boundary between science and non-science. Our survey of different philosophical, legal, and professional attempts to 'define' science has provided us with a good template for this process of disaggregation, and we now turn to the data from our interviewees to develop it.

Notes

1. Scientific Societies Act (1843).
2. I am grateful to David McIlroy for alerting me to the Scientific Societies Act (1843) and its redundancy. See https://hansard.parliament.uk/commons/1960-02-17/debates/35ceb704-cdd7-44e9-a591-4cb0c76442b0/ScientificSocietiesAct1843(Amendment).
3. *Transport Museum Society of Ireland Ltd. v. Registrar of Friendly Societies*, [1999] IEHC 195, High Court of Ireland (https://www.casemine.com/judgement/uk/5da04b1b4653d07dedfd4ce4).

DEFINING SCIENCE   45

4. Guidance: Charitable purposes (https://www.gov.uk/government/publications/charitable-purposes/charitable-purposes).
5. We should not, of course, exaggerate the novelty of this. Science was intensely politicised in the decades after the Second World War, in which its role within the 'military-industrial complex' in the Cold War was scrutinised and criticised. However, the prevalence of climate change scepticism, the anti-vaccination movement, the appropriation of science for a wider range of intellectual disciplines, and the persistence and perhaps growth of pseudoscientific activities have made this a particularly acute problem over recent years.
6. Oreskes, Naomi, *Why Trust Science?* (Princeton, NJ, 2019).
7. Oreskes, *Why Trust Science?*, p. 19.
8. Blakely, Jason, *We Built Reality: How Social Science Infiltrated Culture, Politics, and Power* (Oxford, 2020), p. xxiii.
9. Sokal, Alan D., *The Sokal Hoax: The Sham That Shook the Academy* (Lincoln, 2000)
10. Sokal, Alan D. and Bricmont, Jean, *Fashionable Nonsense: Postmodern Intellectuals' Abuse of Science* (New York, 1998).
11. Gordin, Michael D., *On the Fringe: Where Science Meets Pseudoscience* (Oxford, 2021).
12. Bauer, Henry, 'Science in the 21st century: knowledge monopolies and research cartels', *Journal of Scientific Exploration*, 18(4), 643–660.
13. Ritchie, Stuart, *Science Fictions: The Epidemic of Fraud, Bias, Negligence and Hype in Science* (London, 2020).
14. Ritchie, *Science Fictions*, p. 31.
15. Chevassus-au-Louis, Nicolas, *Fraud in the Lab: The High Stakes of Scientific Research* (Cambridge, MA, 2019).
16. Chevassus-au-Louis, *Fraud*, p. 12.
17. Charles Babbage fingered scientific forgery, along with hoaxing, and data trimming in his *Reflections on the Decline of Science in England and on Some of its Causes*, which was published in 1830, before the word 'scientist' was even invented.
18. Office of Science and Technology Policy (https://ori.hhs.gov/content/chapter-2-research-misconduct-office-science-and-technology-policy).
19. OECD, *Global Science Forum Investigating Research Misconduct Allegations in International Collaborative Research Projects: A Practical Guide, April 2000* (Paris, 2000). https://web-archive.oecd.org/2012-06-14/118158-42770261.pdf.
20. Babbage, Charles, *Reflections on the Decline of Science in England and on Some of its Causes* (London, 1830).
21. Whewell, William, *Philosophy of the Inductive Sciences* (London, 1996 [1840]).
22. Popper, Karl, 'Philosophy of Science: A Personal Report', in Mace, C. A. (ed.) *British Philosophy in Mid-Century* (Crows Nest, New South Wales, 1957)
23. Merton, Robert K., 'The Normative Structure of Science', in Merton, Robert K. (ed.), *The Sociology of Science: Theoretical and Empirical Investigations* (Chicago, 1973).
24. Spencer, Nick, *Magisteria: The Entangled Histories of Science and Religion* (London, 2023), pp. 376–383.
25. 529 F. Supp. 1255 (1982). *Rev. Bill McLean, et al.*, Plaintiffs, *v. The Arkansas Board of Education, et al.*, Defendants. No. LR C 81 322. Section IV(C).
26. Laudan, Larry, 'The Demise of the Demarcation Problem', in Cohen, R.S. and Laudan, L. (eds.), *Physics, Philosophy and Psychoanalysis: Essays in Honor of Adolf Grilnbaum* (Dordrecht, 1983), pp. 111–127.
27. Laudan, 'Demise', p. 124.
28. Laudan, 'Demise', p. 125.
29. *Edwards v. Aguillard*: U.S. Supreme Court Decision. Amicus Curiae Brief (https://www.talkorigins.org/faqs/edwards-v-aguillard/amicus1.html).
30. *Kitzmiller v. Dover Area School District*, 400 F. Supp. 2d 707 (M.D. Pa. 2005) (https://en.wikisource.org/wiki/Kitzmiller_v._Dover_Area_School_District).

46 THE LANDSCAPES OF SCIENCE AND RELIGION

31. The Science Council's definition of science (https://sciencecouncil.org/about-science/our-definition-of-science/#:~:text=Science%20is%20the%20pursuit%20and, Evidence).
32. Pigliucci, Massimo and Boudry, Maarten (eds.), *Philosophy of Pseudoscience: Reconsidering the Demarcation Problem* (Chicago, 2013).
33. Pigliucci and Boudry, *Philosophy of Pseudoscience*, p. 2.
34. Dupré, John, *The Disorder of Things: Metaphysical Foundations of the Disunity of Science* (Cambridge, MA, 1993).
35. Pigliucci and Boudry, *Philosophy of Pseudoscience*, p. 20.
36. Pigliucci and Boudry, *Philosophy of Pseudoscience*, p. 13.
37. Pigliucci and Boudry, *Philosophy of Pseudoscience*, p. 15, quoting Laudan.
38. 'What is this thing we call science? Here's one definition ...' (https://www.theguardian.com/science/blog/2009/mar/03/science-definition-council-francis-bacon#:~:text=Here's%20what%20they've%20come,systematic%20methodology%20based%20on%20evidence.%22).
39. Dawkins, Richard, *The God Delusion* (London, 2006), p. 321.
40. Quoted in De Waal, Frans, *The Bonobo and the Atheist: In Search of Humanism among the Primates* (New York, 2013), p. 98.
41. De Waal, *The Bonobo*, p. 98.
42. Dawkins, *The God Delusion*, p. 321.
43. Duhem, Pierre, *Aim and Structure of Physical Theory* (New York, 1962).
44. Stanford Encyclopedia of Philosophy: Scientific Pluralism (https://plato.stanford.edu/entries/scientific-pluralism/).
45. Ritchie, *Science Fictions*, p. 15.

3
Disaggregating Science

Introduction

All of the 'elite' interviewees—scientists, philosophers, ethicists, theologians, science communicators, journalists—were able to answer the question of what they thought science was. Many did so at length and in considerable detail, answering with clarity and sophistication to repeated questions that probed and tested their ideas. But not many did so willingly. A large proportion, from the outset, expressed significant reservations about the process, and many insisted that it was an inherently artificial and unsatisfactory exercise.[1]

> Even just saying 'science' is problematic ... There are various sciences [but] I don't think there is such a thing as science ... Science is not reducible to a single idea, there are lots of different sciences ... I don't think science is a natural kind or has a core essence. That is, I don't think there's a scientific definition of what science is. (#85, #5, #93, #83)

There were several reasons for this. Some interviewees recognised the imprecision of any boundaries put around science.[2] Others remarked on their porosity[3] and their mutability.[4] 'At any given time, the boundary between science and not science is not fixed. It's always contested and contestable and, in that sense, moving' (#68). What was deemed 'science' had changed over time and even today it boasted no 'necessary and sufficient' conditions or indeed any 'unique characteristics' (#101).

Physics was frequently cited as the paradigmatic science[5] and yet, as a paradigm, its methods (and results) were judged to be 'too distinct' from those of many other 'scientific' disciplines to be helpful in defining the overall category of science (#76).[6] 'If you want to say that particle physics and psychology sit in the same realm of certainty of knowledge you're completely out to lunch' (#85). Moreover, physics itself could be 'unscientific'. One philosopher pointed out that 'lots of theoretical physics is completely detached from observation' (#14), while a number of physicists referenced

48 THE LANDSCAPES OF SCIENCE AND RELIGION

string theory as a discipline that had slipped beyond the realm of what could normally be called science into that of (in *their* words) 'religion' or 'faith'.[7] This, it hardly needs saying, was not intended as a compliment.

In a similar way, just as there were recognised sciences that seemed to fail to meet some fundamental scientific criteria, there were numerous comparable counterexamples of disciplines or 'enquiries' that were not usually labelled science, such as history, journalism, or detective work, but which *did* meet certain 'scientific' criteria pertaining to 'methodology, truth seeking and so forth' (#14).[8] Interviewees were much preoccupied with the liminal space of disciplines that were contestably scientific, citing as examples linguistics (#1), social science (#82), sociology (#13), economics (#13, #95), behavioural science (#13, #56), political science (#29), meta-cosmology (#32), history[9] (#39, #76, #81), literature[10] (#55), psychology (#56), anthropology[11] (#56, #95), mathematics[12] (#57, #60), archaeology[13] (#20, #81), philosophy and formal logic (#66), sports science (#88), engineering (#91, #95), ethnographic research (#95), biblical studies (#100), and psychiatry (#101). Interestingly, legal science was not mentioned, although it presumably falls into the same liminal zone. One brave soul (and not even a theologian) even made a case for theology being included in the category, despite the fact, as we shall come to see, that a number of others defined science specifically *against* religion or theology.[14] Either way, many of these liminal disciplines, if not all, were considered to be sciences,[15] although almost always contestably so.[16]

All this left elite interviewees generally resisting the idea of definitive boundaries for science,[17] or favouring a gradated or hierarchical understanding of the sciences,[18] or denying the principle of taxonomy altogether.[19] That recognised, none of this prevented them from subjecting the concept of 'science' to fruitful disaggregation. People were (often) unwilling to say what exactly science was, but they were prepared to catalogue its various, shifting, constituent elements. They were inclined towards what one respondent called 'pluralistic definitions of what science is' (#88) or what two others (unprompted but gratifyingly) called a 'family resemblance' approach.[20]

Disaggregated 'features' of science

Pigliucci's 'family resemblance' approach to science suggested, if only as a starting point, two constituent dimensions that could serve as a way of disaggregating science: empirical support and theoretical understanding (i.e. internal coherence and knowledge). The previous chapter discerned four normative dimensions—philosophical, methodological, cultural, and social—from within the various philosophical, legal, and official definitions of science it discussed, while also taking into consideration a fifth, 'descriptive' dimension pertaining to the actual practice of science.

Building on and developing this initial disaggregated understanding of science, the expert interviews we conducted oriented us to six dimensions, namely:

1. Method(s)
2. Subject
3. Presuppositions
4. Objectives
5. Values
6. Institutional structure

Each of these encompasses a range of factors, and none is wholly discrete or watertight. All, but especially the last two, incorporate the descriptive, as well as the normative aspect of what science is. Between them, they enable us to build up a broad, composite, and flexible 'family definition' of what science is.

The methods of science

The idea that science could be defined and distinguished by 'the scientific method' was the most widely discussed feature within the interviews. However, at the same time, the idea that there *was* such a thing as *the* scientific method was treated in much the same way as the idea that there was such a thing as science.

People could and did talk about it without complication or reservation,[21] but many questioned the idea that there was such a thing as a single recognisable scientific method—as opposed to a cluster of related scientific methods[22]—and others insisted that there was nothing uniquely *scientific* about the scientific method.[23] In effect, the understanding of the scientific method was a microcosm of the understanding of science itself, in the way in which it was best understood as a fuzzy or family definition.

50 THE LANDSCAPES OF SCIENCE AND RELIGION

For some, in as far as there was a scientific method it was simply the systematisation and formalisation of the kind of interrogative processes that reasonable people engaged in all the time.[24] Defining the scientific method and planting the flag of science foursquare upon it was, it was argued, futile.[25] Nevertheless, describing the methods, or alternatively 'rules' (#4) or 'tools' (#35) or 'practices' (#95), in the plural that were characteristic of science *was* helpful. In effect, methodological considerations were central to the disaggregated definition of science, although it is important to emphasise that these disaggregated methodological definitions were rarely considered to be either (1) unique to science; (2) sufficient to describe 'science'; or (3) necessarily present in any discipline that called itself 'science'. In other words, science had its methods, but (1) it shared them with other disciplines; (2) it needed more than just these methods to be called science; and (3) not all sciences were characterised by the same range of these methods.

What, then, were the different elements—or perhaps 'sub-features'— within this particular methodological feature in the family definition of science? First, the methods of science involve forming 'questions' (#93), 'hypotheses' (#56, #64), 'conceptual models' (#16), and/or 'theories' (#28) about 'reality' (a word deliberately left opaque at this point). Note, 'first' here does not mean that the formation of 'hypothesis' was understood to be the entry point into the methods of science, preceding anything else. The relationship between hypothesis and the gathering of data was something of a chicken-and-egg, with respondents talking of an ongoing dialogue, or dance, between the two. Interestingly, in spite of the distinction between hypothesis and theory drawn in *Edwards v. Aguillard*—namely, that a hypothesis, unlike a theory, tends to be based only on limited data and has a correspondingly restricted assumption of truth—this difference was not widely cited by interviewees, who used these two terms, and indeed the word 'model', interchangeably.[26]

Second, science's hypotheses (and theories) were testable and ideally refutable. The principle—that the methods of science were characterised by the testability of its hypotheses—was widely acknowledged, but consensus broke down when it came to the question of their refutability and, even more so, when it was framed in the language of falsification. A number of interviewees cited Popperian falsification approvingly,[27] while others were considerably more critical.[28] In as far as there was any agreement it was tentative, in as far as falsification was considered to be 'important but not necessary' (#57).

Third, the testing of hypotheses and theories was grounded in a particular set of rules, 'practices', or 'methods' that characterised science, namely the 'ordered' observation of the 'external' or 'natural' world, though both those terms, as we shall see, were inherently contestable (#22, #30, #36, #39, #40). Ideally, though not necessarily, this meant experimentation.

Experimentation was frequently judged as central to science, the quintessential example of isolating and 'controlling' the element of reality that you wanted to observe (#97). It was, in effect, the epitome of the 'structured' or ordered act of observation that was central to science. This structured observation meant that such experiments could and should be repeatable and the results reproducible.[29] Some interviewees pointed out that this characteristic was not unique to science—lots of things, from musical performances to games of sport, could be repeated without qualifying as scientific—whereas others were (painfully) aware that in some apparently solidly scientific disciplines (psychology was the most commonly named), experiments might have been repeatable without their results being replicable.

Fourth, commitment to experimentation was paralleled by commitment to quantification and accuracy. Indeed, it was effectively irrelevant without it. There was close to consensus on the idea that science demanded both accuracy and measurement.[30] The reservations to this, such as they were, pertained not to the idea of accuracy and measurement but to its practicalities. Scientific observations and experiment should entail accurate measurement, and ideally the application of mathematical and statistical techniques to the resulting data in order to achieve greater confidence in the results. However, the reality was that some phenomena were hard to measure accurately, errors crept into measurement, and some data were not amenable to statistical techniques.[31]

These commitments to refutation and replicability, ordered observation and experimentation, and accurate measurement and application of mathematical techniques to resulting data resulted in a fifth key feature, one that was the most frequently mentioned but that also crossed the boundary between the methods of science and the values or ethos of science, namely its 'provisional' approach.

It is important to place this point in its proper context. Some interviewees outlined how this *should* work. 'It doesn't matter if you are the most distinguished professor in the world, some bright young spark can come along and completely overturn everything' (#26). Others, as we shall see in the section on the institutional structure of science, were equally clear that it rarely

52 THE LANDSCAPES OF SCIENCE AND RELIGION

ever worked that way. Just as some interviewees could be dismissive of an uncritical commitment to falsification—the idea that scientists were relentlessly trying to disprove their theories was, they claimed, simply untrue—so were they critical about science's provisionality. The whole point of a rigorously ordered and carefully quantified set of methods was to arrive at conclusions that *were* warranted.

That said, interviewees were also clear that warrant should not be confused with certainty. The very premises of the key features of science's methods—hypothesis, theory, ordered observation, repetition, replication, and measurement—rested on provisionality at every point, albeit the words to describe this point were varied. People spoke of science being 'provisional', 'open', 'tentative', 'sceptical', 'adaptive', 'dynamic', or epistemically humble. The semantic range suggests that any apparently established consensus on this point of science's provisionality is itself slightly vague and perhaps provisional.

Whatever the language used, science's epistemic humility was important. Provisionality could be found at the point of theorisation; indeed, it *was* the point of theorisation. 'A model is just the best guess at what the world would be like, but usually only applies to very specific conditions' (#47). It could be found in the process of observation and in the ensuing relationship between original hypothesis and resulting data. 'It's a sort of never-ending dialogue between data and theory' (#50). And it could obviously be found in the 'conclusions' arrived at and the necessary willingness to challenge them.[32] The result was the conviction that 'all scientific knowledge is provisional' (#21), that science is 'constantly adapting ... [always] fluid and ... open to new discoveries' (#38), and that, taken to its logical conclusion, there was 'no absolute truth in science, because it can always be questioned or challenged or changed or revised, or modified' (#45).

Epistemic humility could also be seen in the cognisance of science's limitations. Needless to say, this was a contentious issue, with some interviewees confident in science's capacity to answer questions, and others dismissing that which science couldn't answer as being non-questions in the first place. Still others, however, were more sanguine about science's abilities, and placed a greater emphasis on its limitations, whether metaphysical,[33] theological,[34] moral,[35] or even material.[36]

As noted, a number of interviewees criticised these convictions as having little bearing on the actual practice of science, and there is certainly a kind of irony in asserting definitively the provisionality of science. Accordingly, many interviewees were keener to characterise the epistemic humility of science in still more humbler terms—as 'a continuing and dynamic process' (#54), 'an iterative process' (#82), 'a cyclical method of both deduction and induction' (#82), or a 'dialectical [approach] that's ... like a relationship or a dialogue' (#68).

They were also keen to delineate science's epistemic humility by means of comparison with other approaches or disciplines that were, apparently, characterised precisely by the opposite approach to the acquisition of knowledge. Thus, science was about 'discoveries, not about tradition' (#38), it was the antithesis of dogma, or, with a nod to its origins in the seventeenth century, an alternative to metaphysics.[37]

Naturally, this comparison also included religion. Occasionally, the comparison was absolute and direct. 'Everything that I can falsify is science. Everything that I can't falsify is religion' (#27). 'One is a belief system, and one is a disbelief system' (#37). At other times, the comparison was acknowledged but also problematised, nodding to the way in which the terminological shifts of the disciplines lay behind this perception.

> I think we define science as that which isn't really religion most of the time, [and] we define religion as that which isn't science. I don't think that's accurate but it's just to show that the concepts, certainly how we use them since the nineteenth century, are completely reliant on one another and entangled together. (#83)

Whether or not interviewees placed science and religion in direct opposition to one another, the overall understanding was of a complex, shifting, aggregated, methodological understanding of science incorporating theorisation, testability, refutation, (perhaps falsification), structured observation (ideally experimentation), accurate quantification, and epistemic humility. This complexity naturally oriented people away from an 'either in or out' approach, and encouraged some to posit a gradated or hierarchical conception of what did and did not qualify as science according to these methodological criteria. Many explicitly or implicitly recognised that physics was judged as occupying the top rung of the ladder, although they did not always concur with that judgement. But for one interviewee at least, it was clear what was on the bottom rung.

> There is an implicit, assumed hierarchy among academics, I think, between who's got hardest science, you know, so the particle physicists would put themselves at the top. And you know [indicating low point with hand], theologians would be there. (#95)

54   THE LANDSCAPES OF SCIENCE AND RELIGION

We will return to this important and contrasting assessment of the 'hardness' of science and religion, or in this case theology, in the chapter on methodology in part II.

The subject of science

The various legal proceedings in the US pertaining to the teaching of 'creation science' and ID, and, in particular, the latter's rejection of solely naturalistic explanations in evolution, have precipitated some admirably direct and clear statements about the proper subject of science.

In their 1998 publication *Teaching about Evolution and the Nature of Science*, the National Academy of Sciences (NAS) stated that 'because science is limited to *explaining the natural world* by means of natural processes, it cannot use supernatural causation in its explanations.'[38] Seven years later, in *Kitzmiller v. Dover Area School District*, the so-called Dover Panda trial, concerning the teaching of ID in Pennsylvania public schools, presiding Judge John Jones III described the 'self-imposed convention of science, which limits inquiry to testable, natural explanations *about the natural world*.'[39]

These observations on the proper subject of science are, of course, closely linked to the discussion in the previous section on the proper method(s) of science. Indeed, Judge Jones's remark continues by highlighting precisely this methodological connection, explaining that science's self-limiting convention of focusing only on the natural world is 'referred to by philosophers as "methodological naturalism" and is sometimes known as the scientific method'.[40] Method and subject are very closely related.

Such a self-limitation of subject was similarly evident among the scientists and philosophers to whom we spoke. *What* science studies emerged as a significant feature science, one that was almost as important as, and inextricably tied up with, the question of the *way* that science studied it. This was repeatedly recognised by interviewees. However, the variety of terms deployed in discussion of this issue (again) underlines how unstable and contestable this issue is. As we will come to note, the idea that science studies 'the natural world' rather begs the question.

We heard that science was a means of understanding 'nature' (#77), or 'the natural world' (#35, #95), or 'the physical world' (#15), or 'the material world' (#79, #81), or more precisely 'how the world functions at a materialistic level' (#34). Alternatively, it was a way of understanding 'the external world' (#33), or 'the observable world' (#79), or the world to which we have 'common access' (#67). Some of these terms were judged in need of greater clarification and parsed further. Thus 'the natural world' was, some interviewees clarified, not to be taken as being in opposition to 'the human world' (#101), because humans 'are natural organisms' (#97). In a similar vein, science studied observable reality, except that strictly speaking that reality was not observable, and certainly not directly observable, meaning that actually science studied 'theoretically observable reality' (#71).

Such refinements were multiplied when interviewees were pressed on this point. We shall return to this discussion in a later chapter in part II, as underlying metaphysical assumptions play a significant role in these debates, constituting one of the major things that people disagree about when they disagree about science and religion. Nevertheless, it is worth noting at this point that almost all the definitions of what science studied could be (and were) problematised.

Thus, science was characterised by the study of the natural (as opposed to the supernatural, rather than as opposed to the human). However, precisely what constituted the category of the 'natural' was far from clear. At the very least, it was a historically contingent category, with events or processes that once appeared non- (or even super) natural now being recognised as perfectly natural, albeit within a more flexible conception of 'natural' than had heretofore been the case. As one philosopher put it, scientists (and philosophers) could be wedded to conceptions of nature that were inflexible, outdated, and essentially wrong.

> One thing that some philosophers do, and other philosophers complain a lot about it, is come to their conception of the nature of reality with this sort of eighteenth century science hat on ... that it's a bunch of little particles bashing around each other. It's basically Newton. And then the supernatural is all the weird stuff that isn't little bits and pieces bashing around each other. That's a terrible view to have about the natural of reality. (#64)

In a similar vein, the 'naturalism' that apparently characterised science was likewise complex and contestable. Naturalism, a number of interviewees observed, 'obviously comes in various forms' (#12), with people distinguishing between methodological and ontological naturalism, or methodological and metaphysical naturalism (#5, #23, #46, #55, #93, #94, #97, #100), or, in one instance, between 'conservative and liberal naturalism' (#17).

There were similar questions around the idea of the material or the physical. One philosopher (not the one quoted above) pointed out that

56 THE LANDSCAPES OF SCIENCE AND RELIGION

'materialism' was commonly associated with a rather 'vulgar account of matter that reality is just made of atoms bumping into each other', and that for this reason, philosophers 'started talking about physicalism ... [because] they thought it sounded less nasty' (#55).[41] However, such socially influenced terminological shifts noted, there were still questions that hung over the terms 'material' and 'physical'.

Several interviewees, physicists in particular, raised questions about how material or physical their subject of study was. 'As a theoretical physicist ... the things I studied way back in my PhD are things that don't exist and that can't exist ... types of universe which we know are not our universe but you can still investigate them mathematically' (#81). The same point was made, even more strongly, about mathematics.

> Many mathematicians are mathematical Platonists, and they think they're exploring a kind of non-natural realm of numbers and triangles and so on. They think it's really out there, mind independent of this reality, but it's not empirically available to us. (#42)

As the word 'available' here indicates, there was a similar problem with the idea that science is necessarily the study of the observable.

> What is really interesting about this is that there is a hardcore naturalism—you stick to what is observed—but when you get to theoretical physics, the hardcore naturalism gets somewhat loosened. We are now hypothesising infinite universes; we have no access to them. (#14)

Even an uncomplicated commitment to 'materialist' or 'physicalist' monism could leave questions hanging over the nature of science. One philosopher pointed out that it was quite possible to hold the view that everything there is in the world ultimately had a physical basis, 'including, for example, mental phenomena, the phenomena of intention, belief, desire, memory, hope, love and so on'. Because the proper subject of science was this physical (or material or natural) world, it should follow that 'everything ... can be explained on that kind of basis', such as the workings of the central nervous or the endocrine system. However, the interviewee insisted, 'it doesn't follow that everything that happens in the universe is reducible to that basis' (#61). Just because the proper subject of science is the physical, it did not mean that everything physical was the proper subject of science.

None of these tergiversations around the words 'natural', 'material', 'physical', 'external', 'observable', and 'available' amounted to a wholesale rejection or undermining of the idea that science was, properly speaking characterised by its subject as well as its method. On the contrary, people were clear that science did have a proper subject of study. However, the problem of defining this was almost as problematic as defining science itself.

The presuppositions of science

The extent to which science's commitment to naturalism is itself an unprovable presupposition is highly contentious. For a philosopher like Alvin Plantinga, there *is* a presupposition to that effect and it is a deeply problematic one: 'there is a deep and serious conflict between naturalism and science'.[42] By contrast, according to Judge Jones's ruling in *Kitzmiller v. Dover Area School District*, 'methodological naturalism is a "ground rule" of science today'. Whether Judge Jones's 'ground rule' is based on warranted and demonstrable reasons, or simply on the basis that naturalistic assumptions have worked so far, is unclear.

What makes this particular issue especially contentious is the way in which advocates of ID theory have used the question of naturalism's sufficiency as a wedge tactic to undermine the adequacy of evolutionary explanations. Thus, William Dembski, a leading advocate of ID who was referenced frequently in *Kitzmiller*, has argued that the principle of methodological naturalism needed to be overturned if ID were to prosper. However philosophically contentious the topic is, it is sociologically and culturally far more so.

Perhaps for that reason, few interviewees thought that science's naturalism (or materialism or physicalism) was a contestable presupposition. The closest anyone came to it was the statement that 'science ... makes assumptions that the world can be explained in terms of matter, but there's no reason why it should be', and this was from an interviewee who was about as far from being an ID advocate as any we spoke to (#75).[43]

That recognised, interviewees did acknowledge that science was predicated on certain 'presuppositions' (again, precise terms varied here) and that these were fundamental, rather than incidental, to its nature. First, there was a commitment to the existence of specific laws[44] and, more broadly, the immutable lawfulness of reality. Science was commonly described as 'the process of trying to uncover laws' (#37) or 'the systematic study of nature to

58   THE LANDSCAPES OF SCIENCE AND RELIGION

extract laws' (#39). Elsewhere, 'pattern' or 'regularity' was the preferred term. Science is 'the discerning of regular patterns within the world' (#16), the attempt 'to come up with theoretical constructs that explain those patterns and predict patterns' (#25).[45]

Whichever term was used, a number of interviews recognised that such laws, patterns, regularities, and lawfulness was presupposed rather than proven by science. In the words of one cosmologist, 'I have no idea where the laws of nature came from and that is actually the surprising thing ... it's a great mystery that anything exists' (#77). One interviewee drew an explicitly theological conclusion from this—'We believe in natural laws and I don't think it makes any sense to believe in laws unless you believe in a law-giver' (#83)—but few others connected the phenomenon with any theological beliefs.

The lawfulness of the universe was not the only presupposition mentioned. A second axiom lay in the conviction that the universe was intelligible or, put another way, that the evolved human brain was cognitively capable of grasping the truth about reality. According to this argument, science '*believe[s]* that the world is amenable to rational explanation' (#7, emphases added). The fact that the world 'could be explained to the satisfaction of human intelligence is not something that we should take for granted'; indeed, to do so 'would be irrational' (#75). Most categorically, 'the fact that our cognitive capabilities would match up to the way the world really is points towards something like a designer that has created us to discover the world' (#83).

Different interviewees drew different religious conclusions from this point. One interviewee (not a religious one) noted the historical–theological connection—'[science] rests on various axioms about the intelligibility of the universe, which are derived—or emerged—originally from theological beliefs, and certainly not from empirical observation' (#2). For others, it was simply an assumption or a mystery. 'There's no particular reason our brains should be matched to understanding all the deep aspects of reality' (#33). Either way, those that mentioned it were clear that science could not work without this presupposition, any more than it could without a commitment to the lawfulness of reality.

A third, final presupposition was, so to speak, deeper than either of these, namely a prior commitment to the existence of the external world and to the reliability of human senses. 'I don't think there are any good philosophical arguments that it is okay to trust our senses, but most of us think ... that all knowledge begins with trusting experience' (#17). For those interviewees who did raise this issue, such a commitment could be termed a properly or legitimately 'basic belief', to be ranked, in the mind of one philosopher, alongside certain moral beliefs or indeed the belief in the existence of other minds. Such basic beliefs are 'foundational in the sense that they seem to be entirely necessary for shared human communication'. It was not that 'they [had] wonderful epistemic warrant'. Rather, it was 'simply [that] we can't get away *without* assuming them ... we can't make any sense of the discourse we're having right now unless we set up these things' (#76, emphases added).

This observation, like the previous two, had (equally contestable) religious implications. One philosopher explicitly stated that they didn't see religious beliefs as falling into the category of being properly basic, whilst another thought that she had 'no problem with saying that belief in God is a properly basic belief' (83). Either way, these comments underline how science was characterised by its commitment to (or, more provocatively, belief in) certain crucial and ultimately unprovable presuppositions without which it could not work.

The objectives of science

The first three features of science—its methods, subject, and presuppositions—are all crucial elements within the overall 'family definition' we are accumulating, but they risk giving the impression of science being a primarily theoretical activity, abstracted from anything that human beings actually do. This issue—'the actual practice of science' for want of a better phrase—was repeatedly raised by interviewees, practising scientists just as much as sociologists of science.

> We continue to allow this kind of myth of science being out there on a cloud almost like a deity, in effect, rarefied beyond the capacity of human beings and science proclaiming on this and science proclaiming on the other. [But we should go] back to the original definition. Science is what scientists do. (#89)

Accordingly, the second group of features—the objectives, values, and institutions of science—draw into the 'family definition' aspects of the empirical reality of science that are important but otherwise in danger of being overlooked.

60 THE LANDSCAPES OF SCIENCE AND RELIGION

This descriptive understanding of science could be in direction tension with the normative one. 'I admit as a scientist that many scientists don't always operate as scientists. They spend a lot of time looking for data that reinforces their own positions, rather than finding ways of testing their own position and being open to alternative views' (#79). A similar sentiment was voiced in numerous other interviews. There was a descriptive–normative tension within the understanding of what science was, just as there is in the science–religion debate.

However, this was not a necessary or inevitable tension, and the fact that science was a socially, culturally, and morally embedded practice did not necessarily undermine its principled dimensions. This can be seen in the ideas that science had certain objectives, goals, or purposes.

Science was characterised as 'a reliable way of finding knowledge about the world' (#10). That reliability was usually established by its ability to make accurate predictions about things. Science was (ideally) predictive. 'Science is most powerful when you've crafted a theory based on data and then that theory takes you to make a prediction where you've not made any measurements before and it turns out to be right' (#50). Science is characterised by its accurate predictions about reality, and, moreover, because scientific knowledge was cumulative as well as predictive, those predictions got more accurate.[46]

This cumulative and increasingly effective prediction located science within a practical framework because, for all that science was in theory the disinterested pursuit of knowledge, it was, in practice, humanity's greatest tool. Precisely because it was successfully predictive, and increasingly so over time, science was usually tied up with human endeavours to control, modify, and improve the material world. It was not simply 'a reliable way of finding knowledge about the world' but, as the same interviewee went on to say, it was something 'that we can make predictions from, that we can use as the basis for technology and engineering and find that it works' (#10). Or, in the words of another, science is a 'construction of models about the world and the testing and extension of those models in a way that leads us to being able to change the world' (#16). Science, in short, could also be characterised by its objectives or goals in the real world.

It is important to make two distinctions here. The first is that not all scientific disciplines could be equally characterised in this way. A number—typically particle physics, cosmology, and mathematics—were, it was judged, more abstract and practically 'disinterested' than others, although that was emphatically not to say that their discoveries could or would not one day be deployed for practical purposes.

Second, even those interviewees who recognised science as having certain objectives still drew a distinction between science and technology, or between science and engineering. This was not to say that engineers were not scientists, but that, just as some scientific disciplines were *less* practically grounded than others, so some, like engineering, were *more*. Thus, in the words of one interviewee, 'the purpose in science in the near term at least is to gain some new insight or understanding, whereas in engineering it is to solve a problem, and, in that sense, it is a practical endeavour' (#91). The objectives (or sometimes 'purposes' or 'goals') of science were more visible in disciplines like engineering than they were in others like evolutionary biology or high-energy particle physics. Once again, it is important to emphasise that within the 'family definition' of science we are pursuing, not all characteristics are universally or equally evident in every area.

The values of science

Provisionality or epistemic humility is, as we have seen, important to the method of science. However, as already mentioned, it would be misleading simply to see such humility as belonging solely to the methodological feature of science. A number of interviewees spoke of it in ethical or cultural terms, as a value inherent in scientific practice. This is important. Science is not simply a process or a method. It is also 'a way of looking at the world' (#1), a 'commitment to internal questioning and to critique' (#68), 'an attitude' (#24), and 'a self-critical aspect' (#57). Indeed, for one interviewee there was no such thing as science so much as 'a generally scientific orientation to acquiring knowledge' (#21).

The importance of humility (also termed 'intellectual ... modest[y]' (#73)) as a value in science was incontestable, albeit often honoured more in the breach than the observance. Recognising this helps to underline how science was as much characterised by its values, implicit and explicit, as its more formalised methods.

Alongside humility, science was exemplified by curiosity. 'Science is an attitude *and* a set of methods. The attitude is open-ended curiosity, wanting to know, wanting to understand, wanting to find out, wanting to observe' (#24). Indeed, in its own way, curiosity was as fundamental to science as any other factor so far discussed. 'Science is driven by curiosity, and it's driven by

62 THE LANDSCAPES OF SCIENCE AND RELIGION

the desire to know something, to replace ignorance with knowledge' (#52). It was no more possible for science to flourish without a tacit allegiance to curiosity than it was without an equally tacit commitment to the lawfulness or intelligibility of the universe.

In addition to humility and curiosity, there was an adherence to what might be termed self-discipline. This was present throughout the scientific process: in the commitment to objective (rather than subjective) explanations; to publicly accessible (rather than privately owned or controlled) phenomena; to a broad, often collaborative, and cumulative approach to acquiring knowledge (rather than narrow, personal intellectual advancement); and to subsequent application of that knowledge for a wider public (rather than private) good. At almost every stage, science presupposed a willingness to withhold or to discipline individual and personal views, desires, and ambitions. Science was the 'attempt to combat our own cognitive biases' (#42), 'to assess ... evidence [in a] dispassionate manner' (#53), and to 'understand ... the world, as it were, from outside' (#75).

Such self-discipline was particularly important in the light of the fact that science is an inherently conflictual enterprise, as multiple respondents underlined, and one in which scientists are (in theory) endlessly engaged in the task of criticising themselves and each other. As one interviewee put it, 'scientists usually winnow understanding from the chaff of confusion by disagreeing with each other' (#58). This is a process that naturally risks provoking resentment and ire and which, therefore, is all the more in need of the virtues attendant on self-discipline, such as openness, patience, attentiveness, and humility. When it works, it works very well. Indeed, as the same interviewee claimed, science can even offer a model for all humanity.

> Occasionally, our disagreements are accompanied by personal rancour, but usually they are friendly. We recognise that we are after the truth, not each other. So, science gives humanity a model for something more than the content of our discoveries—something often sadly lacking in disagreements about politics and religion: how to disagree without being disagreeable. (#58)

That said, it is important to stress that all this is 'when it works', 'in principle', 'in theory', etc. Subsequent accounts of the actual practice of science repeatedly revealed that such rigorous self-discipline was, like epistemic humility, often flouted in practice. The actual practice of science was a good deal more 'human' than this idealised portrayal admits. Moreover, as we shall note presently, the withholding of the self was only beneficial to a certain extent, and scientists were often tacitly expected to bring something of themselves into their work.

Beyond humility, curiosity, and self-discipline, science was (or should be) characterised by a host of other virtues that play quiet and often unrecognised but nonetheless essential roles. There was creativity and imagination. 'Science depends on the imagination to create the hypotheses and decide what's worth pursuing and all those sorts of things' (#67). There was 'intellectual integrity' (#61) and 'moral commitments to truthfulness, to honesty ... hard work ... [and rejection of] negligence and deceitfulness' (#6). And there was even instinct and intuition. As one scientist put it:

> It's very important to acknowledge intuition, subjectivity, a sense of something being the case. And that can happen in science, you just have a feel that something's right or it's not right ... I'm a firm believer that it's all too easy to think that science is all about objectivity ...I think that is such a misunderstanding of science, and all good science. (#45)

These characteristics pick up the point just made, namely that self-denial is only a virtue up to a point. The actual practice of science commonly requires scientists to bring personal traits, such as creativity or intuition, to the table.

Two other points are worth noting, in passing, concerning the importance of values in the understanding of science. Firstly, such moral commitments could also in theory be categorised as part of the presuppositions of science. Indeed, they sometimes were, as one philosopher observed that there are 'a number of moral things ... which clearly are not justifiable in terms of science, but scientists have to proceed with them' (#6). As we have emphasised from the start, the six features within the 'family definition' of science that we are exploring here are not discrete or impermeable, but frequently shade over one another.

Second, in much the same way as scientific method(s) were sometimes explicitly understood in opposition to religion, so were the values of science. As we saw above, at least one respondent viewed the (ideal) practice of science as a model of disagreement that is 'often sadly lacking in disagreements about ... religion' (#58). Others expressed similar views, albeit in different terms. 'Mystery and wonder', said one cosmologist, 'are an important part of science and this redoubles my reason for objecting to dogmatic religion' (#33). 'A crucial feature of science that does distinguish it from religion', claimed another interviewee, 'is that you shouldn't be satisfied with

64   THE LANDSCAPES OF SCIENCE AND RELIGION

the first satisfactory answer you come up with' (#21). By these reckonings, constructive disagreement, an openness to wonder, and a restless critical spirit characterise science in precisely the way they don't characterise (dogmatic) religion.

The institutional structure of science

The sixth and final feature within the 'family definition' of science is its institutional nature. As with the values of science, this is well recognised in the sociology of science, if less prevalent in the popular mind. It is, however, crucial.

At the first level, science is, as already noted, an inherently collaborative and communal enterprise—'a social activity' (#20). Scientists rely on each other, both diachronically and synchronically. Although we're individuals, one physicist said, 'we really all rely on our predecessor's work' (#58). Every single person who engages in science today stands on the shoulders of others. (That this is in profound tension with another element within the family definition of science, namely its willingness to question everything, will be clear and is something to which we shall return.) Similarly, each scientist stands alongside their peers. 'It is essentially a [collective] human activity, more brains than just our own ... it's the closest thing we have to a group mind' (#58). Both across time and across borders, science is a shared activity.

So significant is this that people claimed that the 'solitary scientist' was an oxymoron. Although some clearly tend towards more solitary work, none works in complete isolation.

> The thing that we call science today is a collaborative social exercise through which we are able to establish agreed knowledge ... The knowledge is only as good as the inclusivity of that collaboration ... A person doing science in their shed who never reads any articles, who never publishes any articles, never submits anything to peer review, they're not scientists, they're a crank. (#96)

This quotation underlines not only how essential collaboration is to the practice of science, but also, in its reference to 'agreed knowledge' and 'peer review', how such collaboration is thoroughly institutionalised. The communality of endeavour and attainment is not casual or informal but officially structured and recognised.

This institutionalisation manifests itself in (at least) two ways. The first concerns the specific institutions or organisations within which scientists work. This is typically, but by no means solely, universities. 'Scientists working within universities are working to certain norms and practices and goals that are imposed by those organisations and obviously the REF [Research Excellence Framework] and everything like that has huge impact on how science is conducted and how scientists identify' (#81).

The second is the sense of belonging to a wider community of scientists, which institutionalises people through norms and practices around things like article writing, peer reviews, journals, and conferences, 'the things that scientists do collectively across the whole of science that binds them together no matter how disparate their actual practices in doing their science' (#81). Through either and usually both of these ways, science operates as an institution.

Such institutionalisation has grown in significance over the years, as science has become bigger and more expensive. 'What is modern [about science] is having a huge, organised community with certain sorts of funding and authority and social position' (#67).[47] These developments have further undermined 'the traditional view of the scientist is a single genius in a room' (#54).

The institutionalisation manifests itself particularly with regards to language and structures of authority. First, language. In theory, science should be open, 'accessible in principle to third parties' (#2), 'a completely democratised route to knowledge' (#8), committed to 'public accessibility' (#46). In reality, its institutionalisation gives it a linguistic identity which is often inaccessible or incomprehensible to those outside. Science is 'the framework that allows communication between different claims about reality', 'a common language' that allows cross-cultural communication providing it is between those who are within the institution (in the second sense) of science (#79).

One interviewee remarked that 'if I talk about science in a particular context, you know exactly what I mean. But once you try and do this in a context-free context ... it becomes enormously difficult to define what science is' (#88). It is, to return to Wittgenstein, a language game only comprehensible within a particular form of life, a form with which non-scientists are largely unfamiliar. In a similar vein, another interviewee referenced the

66 THE LANDSCAPES OF SCIENCE AND RELIGION

British comedians Flanders and Swann saying, 'you have to learn the language of scientists because they don't understand anything else, the poor dears', explaining that 'you have to become enculturated within a particular language, within a particular methodology' in order to understand it (#89).

Second: structures of authority. If, as one sociologist put it, 'science is what scientists do' (#74), the following question—who, then, is a scientist?—can only be answered with reference to institutions. Scientists are people who are formally recognised as scientists. Science is 'an institutionalised way of knowing' (#81). The 'order, method, and structure' of science 'are agreed on by a community of knowledgeable peers' (#22). 'Science is a profession' (#29) and thus in some way dependent on formalised structures of professional recognition. 'Science is something that's practised by those who have a certain kind of expertise which we recognise as a function of their training, their status within certain institutions' (#68). Progress in scientific knowledge is made through the process of 'peer review', the clue being in the name (#42, #95).

This recognition of the inherently institutional nature of science, complete with linguistic codes and structures of authority, was, not surprisingly, compared to religion by a number of respondents, usually those with some familiarity with religion. In this instance, unlike those above, the comparison pointed to the similarity rather than the difference. Because science is 'an actual community of people in a shared endeavour', it develops 'its own methods, it has its own rituals, it has its own hierarchies, it has its own reward system ... hierarchies and authority structures'. This made it 'not quite [a] faith but ... in some ways religion-like' (#67). Elsewhere, one respondent (correctly) pointed out that it was this almost competitive institutionalisation of science in the nineteenth century that lay at the heart of the (historical) conflict between the two.

> I do think the professionalising of science and the professionalising of religion have a lot to do with the conflict between religion and science ... the priesthood of science versus the priesthood of religion. Rival priesthoods can be vicious in their conflicts with each other. (#73)

Again, the role that the institutionalisation of science plays within the wider perceived conflict between science and religion is something to which we will return in part II. For the time being, we simply want to draw attention to the deep significance of science as an institution as part of the wider 'family definition' we have been assembling here.

Conclusion: outliers and dissensions

These six features or dimensions—the method(s), the subject, the presuppositions, the objectives, the values, and the institutional structure of science—are intended to capture the variety of elements that exist within a 'family definition' of science.

Such an approach emphasises that a discipline does not need to encompass every one of these features—let alone every element within each of the six features—to warrant being called 'science'. It further emphasises that other disciplines or areas of life may exhibit some of these characteristics without necessarily warranting the label 'science'. To repeat the premise of this approach, a Wittgensteinian family resemblance relieves us of the ultimately futile attempt to define something by identifying necessary or sufficient conditions or by locating a firm and impermeable boundary along which we can draw our taxonomic lines. Rather, it offers a range of factors by means of which we can assess whether an activity or discipline looks more or less like science. In effect, it eschews definitive and absolute definitions, the perennial problem of what Richard Dawkins has called the 'discontinuous mind', in favour of a more aggregative, gradated, tentative, and flexible understanding.

As a way of underlining this point, it is worth concluding this discussion with two observations. The first is that the disaggregated definition we have offered in this chapter resulted from the frequent concurrence of answers. In effect, interviewees often converged on elements and sub-elements in such a way that suggested that the resulting definition was not based on individual quirks but on commonly recognised features within the complex phenomenon that is 'science'.

'Commonly recognised' does not mean universally recognised, however, and there were interviewees who articulated altogether more eccentric or less familiar definitions of science. For one, science is 'a way of bringing meaning to our lives' (#8). For a second, it was a process for 'bringing what is far nearby by keeping some things constant' (#90). For a third, 'it's a way of looking at the world and asking how did that happen, why did that happen?' (#43). A family resemblance approach to understanding science cannot definitively say that these definitions are incorrect, simply because such an approach does not claim that any definition is definitively correct. What it does say, however, is that these definitions tend to lie some distance from the centre of gravity when it comes to understanding science. They bear only a faint resemblance and are at best, so to speak, distant family members.

68 THE LANDSCAPES OF SCIENCE AND RELIGION

The second point is that because there is no definitive in-or-out element to this definition, it is possible (indeed probable) that someone somewhere will object to some of the elements we have outlined above. The interviews demonstrated that this was undoubtedly the case. Thus, to take just a few examples: science demands observation, except that 'lots of science isn't like that. Lots of theoretical physics is completely detached from observation ... They hypothesise dimensions we can't observe' (#14). Science is an empirical and ideally experimental process, except that 'within evolutionary theory there're all sorts of speculative propositions which make coherent sense but are not based upon empirical experimentation' (#6). Science is about the material or physical world, except that quantum theorists,[48] anthropologists,[49] and mathematicians[50] often find themselves excluded by this understanding.

Science proceeds by falsification, except 'that can only be said by someone who has never published their own theory in science because the last thing that a scientist wants to do is publish a theory and then immediately start to look for falsifying evidence from what you have just published' (#16). Science demands objectivity, except that, as we have seen, it also requires practitioners to bring their own creativity and judgement into the process, and that elsewhere, in some scientific disciplines, objectivity may simply be a chimera. '[In] social science ... because they are more concerned with humans, this idea of objectivity is more difficult in practice to attain and some people would anyway say, not even so desirable' (#11). Science should be repeatable and reproducible, except that 'a lot of science doesn't work like that at all ... a lot of science isn't repeatable' (#14). Science requires quantification, except that 'a good deal of science, possibly the most of science, still has to be discussed through language rather than through equations ... you have numbers of variables, and many of them can't be quantified' (#66).

For any absolute and definitive attempts at definitions, such criticisms would constitute insuperable barriers. They do not for 'family resemblance' definitions, however. Observation, experimentation, and falsification may be central features in science, but an activity may still legitimately be labelled scientific if it fails to demonstrate these but does, say, feature theorisation, mathematisation, and testability. Complexity is the price of accuracy and nuance. A Wittgensteinian family resemblance offers the best means of grasping what science is, but it comes at the cost of abandoning hopes of a simple, irrefutable, and definitive definition. It is a price worth paying, however, as not only does it help us better grasp what science is,

DISAGGREGATING SCIENCE 69

but also it provides the foundation for a more accurate, sophisticated, and fruitful dialogue between science and 'religion', to which we now turn.

Notes

1. 'That's such a contentious question I don't even know where to start with it ... the best one can do is some kind of loose criteria that covers most scientific enquiry, but then there are always going to be exceptions and it's all going to be a bit messy around the edges.' (#64) 'There is no single thing that is science.' (#96)
2. 'It is difficult to put boundaries on that and say where it begins and ends.' (#29)
3. 'I most certainly see it as a kind of institution that involves a set of practices, but there is a kind of porous boundary to that institution and who is involved in conducting those practices.' (#30)
4. 'The nature of science does change over the history of science, it's changed over the last few hundred years.' (#32) 'For most of history until the nineteenth century, it had a very general meaning, to mean enquiry into the gathering of data, facts, of information.' (#61) 'You can't generalise about it because different sciences are different. They're historically different and they have very different histories, different genealogies and different methodologies and agendas that go with those different histories ... the attempt to bring them altogether under one umbrella called "science" in the nineteenth century, was a rather artificial project.' (#97)
5. 'The paradigm science is physics ... any of its laws should be exposed to falsification and to testing ... And it has delivered in spades a greater understanding of the world; a broadening of our consciousness ... it is a predictive certainty as well, which is the important thing.' (#12)
6. For example, 'some people even look on evolutionary theory as highly dubious as it doesn't measure up to the demanding standards of physics' (#57).
7. 'I call string theorists basically religious. They are. Because they can't falsify it right now.' (#27) 'There is this whole idea of super string theory. Which even its most happy advocates agree can't ever be tested. That is a faith.' (#29)
8. *In extremis*, this could even be applied to ordinary knowledge about the world, e.g. 'I don't think that there is any inherent distinction between how we know ... the age of the earth and how we know what my street address is. Yet the age of the earth really is science, and my street address is not science' (#46).
9. 'History is a nice example because I think it sits pretty close to certain domains of science, history is or should be something that is referring to the empirical record, it's just not in the business to make predictions, which I think is why we put it on that side ... historians are also interested in learning for the future what we can from the past, so even that is not a neat distinction.' (#76) 'Historians might want to call themselves scientists and certainly there was a time when they would be claiming that term. They might be more cautious about doing that now, so what they have in common is that it's systematic and it's evidence-based.' (#81)
10. 'I actually think a lot of literature should be counted as science too. That's more complicated. Literature isn't just in the business of telling you how things are, but quite a lot of literature is in the business of doing that and I think it's good just to extend and get things right.' (#55)
11. 'As an anthropologist, was I a scientist? Some people would say I was because I used rigorous techniques. But I suppose I was more interested in culture and myth and story and narrative. Is that science? Some people say the scientific method can drift into that. But on the whole, if you are looking at that, I think most people would say that is outside the scientific realm.' (#56)

70 THE LANDSCAPES OF SCIENCE AND RELIGION

12. 'If you are happy to relax your definition, then probably I [as a mathematician] may be a pre-scientist, and the stuff I work on maybe one day could have something to do with the physical world.' (#60)

13. 'I guess what [history] lacks compared to natural science is that intervention with material reality and that's where archaeology then straddles that divide.' (#81)

14. 'It's simply a way of asking questions. In that sense, it's not different to theology. Theology relies on the different sources of evidence to inform its arguments and discussions. It might benefit by adopting the methods of science to speed up its processes of discussion maybe.' (#51)

15. 'To me the difference between the sciences proper and if you like the humanities is not really whether or not they're doing science, it's just the technical, it's a quantitative difference in the technical way they go about hypothesis testing.' (#51)

16. 'I don't see any big reason for distinguishing between history and science or sociology and science, psychology. I guess a lot of psychologists want to sell to scientists, perhaps not so many sociologists.' (#55) 'I tend to include human sciences in there as well; so economics, sociology, behavioural science which I think are all becoming increasingly scientific as they use a scientific method, they use theory hypothesis, falsification and that sort of process.' (#13) By comparison: 'I think of history as being an evidence-based subject, but it's not scientific.' (#35)

17. 'That's always going to be dialectical and that's why it's not a boundary, it's like a relationship or a dialogue.' (#68)

18. 'There is a contemporary view of science, which is a really reductive [and] naturalistic. Physics is at the core ... and the soft sciences are around that, and then the social sciences ... as human beings come into the mix ... psychology and economics—half in the camp and half not.' (#14)

19. 'I reject taxonomy ... definitions in almost all examples are restrictive and not enlightening.' (#35).

20. 'I think there are some family resemblances between most of the things we call science.' (#97) 'You might even want to move in a family resemblance direction. People do that for science as well, but the important point is there's no reductive definition here.' (#93)

21. 'It's an experimental method to observe phenomena and develop rules and practises around the observation of that phenomena.' (#71)

22. 'I don't believe that there is a scientific method, but I think there's a collection of scientific practices that have proved themselves to be a reliable way of finding knowledge about the world.' (#10) 'There's no such thing as the scientific method. There are scientific methods, but there's no scientific method, which is the kind of lie we tell kids at school.' (#15) 'I don't really think there is such a thing as universal scientific method. I think the methods we use in biology are too distinct from the methods we use in physics.' (#76) And many similar remarks.

23. 'The scientific method is something which is hyped and which I think is irrelevant. There is nothing very different from what scientists do from what a detective or lawyer does in following up clues, looking for evidence, and testing the reliability.' (#33) 'One way of thinking about science is in terms of the methodology ... the problem with that is that there is all kinds of enquiries that would meet that ... [such as] what a historian is doing.' (#14) 'I think people fetish on scientific method and I think it could come up something like: have hypothesis, test hypothesis. But I don't think there's something particularly reified about the formulation that people say, "Oh, that's the scientific method".' (#78)

24. 'The systematization of a kind of rational empirical methodology we use in everyday life anyway ... we do often formulize and systematize more than we do in ordinary enquiry.' (#76)

25. 'Philosophers of science often try and define science, scientific method. I don't think there's any mileage in doing that.' (#55)

26. There was one interviewee who explicitly drew attention to the difference, then failed to go on to define it. 'The difference between a theory and a hypothesis is telling. I think and

DISAGGREGATING SCIENCE   71

I saw there was some conservative MP a few weeks ago complaining, "Oh the scientists keep changing their minds about face masks", I think it was. We've got to get these scientists under control. That's the whole bloody point. Isn't this supposed to be something that changes as new evidence comes in?' (#40)

27. 'I would basically buy into the classic Popperian accounts that testability is what defines science, so if something is subject to empirical testing and falsification, then it counts as science. If not, then probably not.' (#28)

28. 'People often bang on about Popper and Falsification as being the be all and end all of science. But if you actually look at the history of how science has panned out ... often we haven't necessarily followed falsification strictly.' (#40)

29. 'Science is the use of repeated experiments to interrogate the way in which the material realities of the world ... function.' (#63)

30. 'It's something that you can measure.' (#8) 'It tends to be quantitative.' (#12) 'Usually there is a quantifiable dimension.' (#59) 'The ability to measure, the ability to determine accuracy.' (#69) And many similar remarks.

31. 'We need to make measurements and we have to be aware of the limitations of measurements, errors in particular.' (#47)

32. 'It reaches conclusions that are always tentative but are always open to revision, not to say falsification.' (#28)

33. 'Science is basically telling us how things in the universe work at the level of physical structures, I guess. In that sense, I don't think it can tell us anything about ultimate reality in a metaphysical sense, I think, but then I think nothing can, so that's not science's loss.' (#3)

34. 'It can't actually answer questions to do with things like God. It's a category error to suppose that you can find God, measure God, and sort of exclude the possibility or include the possibility.' (#76)

35. 'I think Einstein was right when he said that you can talk about the ethical foundations of science, but you cannot talk about the scientific foundations of ethics.' (#94)

36. 'What is the job of science? To construct a theory that can account for all the data and observation and experiments. Once you have done that, job done. But I think if you were strictly following that, you would have no reason to believe in consciousness.' (#90)

37. 'When philosophers like Husserl are encouraging us to take a scientific viewpoint, they're not encouraging scientism and they're not talking about natural science or empirical science in that context. They're making the point that we have to ensure that our methods yield conclusions that can be justified and tested and it's really a warning against metaphysics.' (#93)

38. National Academy of Sciences, *Teaching about Evolution and the Nature of Science* (Washington, DC, 1998), p. 127; emphases added.

39. *Kitzmiller v. Dover Area School District* (2005); emphases added.

40. *Kitzmiller v. Dover Area School District*; emphases added.

41. As far as this interviewee was concerned, there wasn't a principled difference between the two terms.

42. Plantinga, Alvin, *Where the Conflict Really Lies: Science, Religion, and Naturalism* (Oxford, 2011), p. xiv. He goes on to explain: 'Taking naturalism to include materialism with respect to human beings ... it is improbable, give naturalism and evolution, that our cognitive faculties are reliable ... but then a naturalist who accepts current evolutionary theory has a defeater for the proposition that our faculties are reliable'.

43. To the best of our knowledge none of the interviewees propounded any form of ID, as that phrase is commonly understood in science and religion discourse, even as some of them were committed to the idea of an intelligent designer.

44. 'Why should you have a universe in which the speed of light is fixed? Why should you have the speed of light be fixed at the value it was fixed at? This is stuff we have no answer for.' (#43)

45. In a similar vein, science is 'trying to see patterns and regularities' (#33) or it is the 'search for kind of big patterns for generalisation, for law, for underlying laws' (#74).

72 THE LANDSCAPES OF SCIENCE AND RELIGION

46. Science is 'the collective knowledge which is being built up by the community of scientists over the years' (#16).
47. In a similar vein, 'science requires laboratories, it requires equipment ... sometimes it requires rockets and very expensive things, so it's got to be an institution' (#54).
48. 'In the old classical view, science was dealing with the material world connected with ... solid objects, tables and things like that. But we know that the classical view got overthrown at the turn of the twentieth century, in particular with quantum theory ... it's surprising to see how many people still think that science is associated with materialism.' (#32)
49. 'As an anthropologist, was I a scientist? Some people would say I was because I used rigorous techniques. But I suppose I was more interested in culture and myth and story and narrative. Is that science? Some people say the scientific method can drift into that. But on the whole, if you are looking at that, I think most people would say that is outside the scientific realm.' (#56)
50. 'In my research [as a mathematician], I am completely uninterested in how the physical world works.' (#60)

4
Defining Religion

Introduction

'Most books on the philosophy of religion try to begin with a precise definition of what its essence consists of.' So William James began the second of his famous Gifford Lectures on religious experience. Definitions, however, he went on to say, are a rather hubristic endeavour. The fact that there were so many—and James was lecturing in 1901–02—and that they were so different from one another should be enough to 'prove that the word "religion" cannot stand for any single principle or essence'.[1]

James went on to invite his audience to take religion as 'the feelings, acts and experiences of individual men in their solitude, so far as they apprehend themselves to stand in relation to whatever they may consider the divine', but he conceded that this was an 'arbitrary' definition, and that we should be prepared to admit that 'we may very likely find no one essence, but *many characters which may alternately be equally important* to religion'.[2]

This may be read as an early prompt towards a family resemblance approach to defining religion: no one definition but a group of characteristics of varied importance. It is an approach that is more familiar for religion than for science and, given the intractable tergiversations around the meaning of the term, one that is even more necessary.

This chapter will describe, necessarily briefly, various attempts to formally define religion, legally and philosophically, over the last hundred years or so, and harvest the resulting set of resemblances, before proceeding, in the following chapter, to our own Wittgensteinian 'family definition' approach to this question.

Defining religion in English law

The need legally to define religion is, on the surface of it, rather more urgent than the need to define science. Whether or not 'religion is natural whereas science is not', as the title of one recent book puts it, it is certainly more

74 THE LANDSCAPES OF SCIENCE AND RELIGION

widespread—deeply felt and socially salient wherever human beings have lived.[3] Given the way in which law demarcates and allocates rights and duties, privileges and exemptions, and legal and fiscal advantages and burdens, knowing whether something should be classified as a religion—with all that means for the position of religion in society—is extremely important, with ramifications for pretty much everyone (including, by definition, the non-religious).

In spite (or perhaps because) of this, attempts to define religion in English law are a relatively recent affair. During England's immediate post-Reformation history, the only lawful religion was the church established by law, and even as the bounds of religious toleration extended, very slowly, from the late seventeenth century, because, unlike many other European countries, England never developed a compulsory scheme for the registration of religious groups, and the need formally to define religion was attenuated.

This has changed in the last 50 years or so. A case brought before the Court of Appeal in 1970, *R v. Registrar General, ex parte Segerdal*, concerning whether a Church of Scientology chapel could be registered as a meeting place for religious worship, rejected the claim, under the Places of Worship Registration Act 1855. It did so on the grounds that worship was characterised by 'at least [some form] of ... submission to the object worshipped, veneration of that object, praise, thanksgiving, prayer or intercession', none of which, the Court held, was evident in the activities of Scientology.[4] Although not technically adjudicating on the definition of religion, the Court's foregrounding of worship influenced a subsequent case, *Re South Place Ethical Society, Barralet v. AG*, which was concerned whether the South Place Ethical Society (SPES) qualified for charitable status under 'the advancement of religion'. The SPES existed for the 'study and dissemination of ethical principles and the cultivation of a rational religious sentiment', but by the logic laid down in *Segerdal* it did not qualify as a religion because, the judge (Lord Toulson) concluded, it witnessed '[no] worship in the sense which worship is an attribute of religion'.[5]

The inadequacy of these definitions was already in evidence even as they were passed, as both judgements admitted Buddhism was an 'exception' to their rulings. *Segerdal* was eventually overturned by the Supreme Court in 2013. In the intervening time, however, the Human Rights Act 1998 incorporated Article 9 of the European Convention on Human Rights into domestic law, and this broadened the understanding of religion, by pairing it with 'thought [and] conscience' on one side,[6] and with 'belief', in terms of manifestation, on the other.[7] The result was the significant expansion of a heretofore narrow category as the courts were called upon to adjudicate on whether a wide range of 'beliefs' merited protection under the Act. Courts on the continent considered claims concerning Scientology, druidism, pacifism, communism, Nazism, atheism, pro-life, Divine Light Zentrum, and the Moon Sect, while those in England heard cases on hunting, climate change, ethical veganism, spiritualism and psychic powers, anti-fox hunting, belief in the virtue of public service broadcasting, and (secular) humanism.

Nobody was pretending that these activities qualified as religions. However, that particular question no longer mattered. The way in which religion had been paired with belief in Article 9 effectively bypassed and rendered immaterial the question of whether something was a religion. Even if it weren't, it might be the kind of 'belief' that merited similar protection. Moreover, when the Equality Act 2006 removed the word 'similar' from the previous regulations—which had defined 'religion or belief' as meaning 'any religion, religious belief, or *similar* philosophical belief'—the delineation of religion became a complete irrelevance.[8] Even were an activity to fail to qualify as a religion, as Scientology had in *Segerdal*, it would almost certainly qualify as a 'philosophical belief'.

This was not the end of the matter, however, as the Charities Act 2006, which clarified and codified centuries of charity law, did define religion without pairing it with belief. Moreover, it specified that 'religion' included 'a religion which involves belief in more than one god, and a religion which does not involve belief in a god', in such a way as would have opened the classification to the previously excluded Scientology and Buddhism.[9] The Charity Commission laid out its logic in an accompanying document on the 'Analysis of the law underpinning The Advancement of Religion for the Public Benefit', which outlined four broad elements of the definition.

First, there was 'belief in a supreme being', although the Commission was clear that it would not 'be proper to specify the nature of that supreme being or to require it to be analogous to the deity or supreme being of a particular religion'.[10] In a subsequent decision on an application for charitable status by the Gnostic Society, this was elaborated as 'belief in a god (or gods) or goddess (or goddesses), or supreme being, or divine or transcendental being or entity or spiritual principle, which is the object or focus of the religion (referred to ... as "supreme being or entity")'—which seemed to cover most bases here.[11]

Second, there was worship of that 'being or entity', meaning acts of 'submission, veneration, praise, thanksgiving, prayer or intercession'. This

76 THE LANDSCAPES OF SCIENCE AND RELIGION

was an intentionally narrower conception than the 'teaching, practice and observance' outlined in Human Rights legislation, and was glossed over in the Gnostic Centre decision as 'a relationship between the believer and the supreme being or entity by showing worship of, reverence for or veneration of the supreme being or entity'.[12]

Third, a religion needed to exhibit 'a certain level of cogency, seriousness, coherence and importance'.[13] And fourth it had to 'have a clear structure and belief system', with 'a positive nature, impacting beneficially on the community', and 'tend directly or indirectly to the moral and spiritual improvement of the public'.[14]

In spite of the move away from the need to define religion in English law (with the introduction of the category 'philosophical belief'), and the dropping of the stipulation that such beliefs needed to be 'similar' to religion, *and* in spite of the detail of the Charity Commission's reasoning concerning their definition of religion, there remained considerable legal confusion in this matter.

When the Church of Scientology returned to court, in 2013 (*R (on the Application of Hodkin) v. Registrar General of Births, Deaths and Marriages*), to challenge *Segerdal*, the judge overturned the decision on the grounds that the understanding of religion 'in today's society' was broader than it had been in 1970. The phrase 'place of meeting for religious worship' from the 1855 Act had to be interpreted in 'accordance with contemporary understanding of religion and not by reference to the culture of 1855'.[15] Similarly, the way in which *Segerdal* had already had to make an exception for Buddhism strongly suggested that 'unless there [was] some compelling contextual reason for holding otherwise, religion should not be confined to religions which recognise a supreme deity'.[16] The meaning of 'worship' given in *Segerdal* was also 'unduly narrow', and the judge approvingly cited the *Concise Oxford English Dictionary*'s definition of 'worship' as including *both* 'the feeling or expression of reverence and adoration of a deity' *and* 'religious rites and ceremonies'.[17]

The effect of all this was not only to overturn *Segerdal*, which was no surprise, but also to drive a coach and horses through the Charity Commission's definition and analysis of religion, which had placed a premium not only on the relevance of a(n albeit rather undefined) supreme deity but also on the interpretation of worship as a form of 'submission, veneration, praise, thanksgiving, prayer or intercession'. Consequently, and in his summary, the judge wrote that he would *describe not define*[18] religion as:

> a spiritual or non-secular belief system, held by a group of adherents, which claims to explain mankind's place in the universe and relationship

with the infinite, and to teach its adherents how they are to live their lives in conformity with the spiritual understanding associated with the belief system.[19]

As a definition it was long a long way from *Segerdal* or *South Place Ethical Society*, as it inevitably would be, but also a long way from the Charity Commission definition. Indeed, in many ways, it was closer to the idea of 'religion ... or similar philosophical belief' of the Employment Equality (Religion or Belief) Regulations 2003, distinguished only by its double use of the word 'spiritual'.

This was underlined by the fact that Lord Toulson went on to explain that he had deliberately eschewed the word 'supernatural' in his definition 'because it is a loaded word which can carry a variety of connotations', and that by 'spiritual or non-secular' he had meant 'a belief system which goes beyond that which can be perceived by the senses or ascertained by the application of science'. This, of course, could apply to almost any philosophical system (including, ironically, scientism itself), though it is nonetheless noteworthy that the judge's expansive 'definition' of religion still somehow managed to pivot on a kind of opposition to science.

In short, after *Hodkin*, the definition of religion in English law remains unclear. At one point in his judgement, Lord Toulson observed that the experience across the common law world had shown the 'pitfalls of attempting to attach a narrowly circumscribed meaning to the word [religion]'. The different contexts in which the issue arose, the variety of world religions, the emergence of new religious practices, and the changing cultural context for such evaluations all combined to render legal definitions problematic. As a result, 'there has never been a universal legal definition of religion in English law'.[20] After his own judgement, with its insistence that whatever idea he alighted on was merely a description, there is still no such definition.

Defining religion in American law

Lord Toulson's attempt to describe religion drew on decisions in other jurisdictions, including the US. For all that the US was founded in opposition to the established religious settlement(s) of Great Britain at the time, the *understanding* of religion in American law was, naturally, the same.

Article VI of the Constitution declared that 'no religious Test shall ever be required as a Qualification to any Office or public Trust', and the Bill of Rights, ratified four years later, declared that 'Congress shall make no law respecting an establishment of religion, or prohibiting the free exercise

78 THE LANDSCAPES OF SCIENCE AND RELIGION

thereof'. In neither case was the category of religion defined and, in as far as it is possible to tell, the Founding Fathers understood it in terms that would have been entirely familiar in the Old World. James Madison, for example, characterised it as 'the duty which we owe to our Creator, and the manner of discharging it'.[21]

It was nearly a hundred years before the Supreme Court was called upon formally to define the term, and when it did, in *Davis v. Beason* in 1890, it did so in terms that would have been entirely familiar to the Founding Fathers and to English law at the time, namely having 'reference to one's view of his relations to his Creator, and to the obligations they impose of reverence for his being and character, and of obedience to his will'.[22] The Court maintained this tight definition when it next returned to the issue in 1931, in a case concerning a Canadian theologian who wished to be naturalised but said he would swear the Oath of Allegiance only on the basis that he would take up arms only in (what he considered to be) a just war. In this instance, the Court defined 'the essence of religion [as the] belief in a relation to God involving duties superior to those arising from any human relation'.[23]

Thirteen years later, the Court, without explanation, significantly widened its definition in claiming that religious freedom included 'the right to maintain theories of life and of death and of the hereafter which are rank heresy to followers of the orthodox faiths'.[24] It maintained this more expansive definition in 1953, when it intimated that any attempt to reduce religion to theism was unacceptable. 'It is no business of courts to say that what is a religious practice or activity for one group is not religion under the protection of the First Amendment'.[25]

This broader understanding of religion was made explicit and cemented in a landmark ruling in 1961, when the Court declared that religion did not necessitate belief in God, and remarked in a footnote that 'among religions in this country which do not teach what would generally be considered a belief in the existence of God are Buddhism, Taoism, Ethical Culture, Secular Humanism and others'.[26] This ruling did not provide any guidelines for determining which beliefs and practices might then qualify as religious, however, and it was in another case, *United States v. Seeger*, concerning exemption from military service, that the Court (began to) clarify its definition. This ruling proposed a test for determining when a belief was a religious belief, namely whether it 'is sincere and meaningful [and] occupies a place in the life of its possessor parallel to that filled by the orthodox belief in God'.[27] A subsequent case, also concerning exemption from military

DEFINING RELIGION 79

service, further developed this expansive approach to religion, bracketing 'moral [and] ethical ... principles' alongside religious ones, and in contrast to beliefs that rest 'solely upon considerations of policy, pragmatism, or expediency.'[28]

Circuit Courts have built upon the Supreme Court's foundations in broadening the definition of religion. One of the most thorough, referenced by Lord Toulson, was the case of *Malnak v. Yogi* in 1979, in which the question before the court was whether 'the *Science* of Creative Intelligence–Transcendental Meditation', as taught in the New Jersey public schools, constituted a religion.[29] (A better example of quite how messy and confusing the categories of science and religion can be would be hard to find!)

In this case, the judge formally distanced the law from earlier, narrower conceptions of religion, saying that the 'theistic formulation presumed to be applicable in the late nineteenth century cases is no longer sustainable'. In its stead, and recognising that the Supreme Court's rulings still left a lot of detail to be desired, the judge outlined 'three useful indicia' that he considered to be 'basic to ... traditional religions', in line with 'the values that undergird the first amendment', but also applicable beyond the narrow conception of religion that had, until quite recently, been dominant in American legal rulings.

The first and most important of these was 'the "ultimate" nature of the ideas' involved, which pertained to 'the meaning of life and death, man's role in the Universe, [and] the proper moral code of right and wrong'. This alone was not enough, however, as 'certain isolated answers to ultimate questions', such as the theory of the Big Bang, were not necessarily religious 'because they lack the element of comprehensiveness'. This comprehensiveness—meaning that, properly speaking, a religion also 'lays claim to an ultimate and comprehensive "truth"'—constituted the second of the three indicia. Finally, the judge wrote that religions were characterised by 'formal, external, or surface signs', such as 'formal services, ceremonial functions, the existence of clergy, and a structural organization'. He acknowledged that although a religion could exist without these, they did tend to provide evidence that a group or belief system was religious.[30]

Another case, around the same time as *Malnak v. Yogi*, placed a similar emphasis on the importance of 'ultimate concerns' within the definition of religion. *International Society for Krishna Consciousness, Inc., v. Barber* judged whether Krishna Consciousness qualified as a religion and declared that the test from *United States v. Seeger*, which treats an individual's 'ultimate concern', whatever it may be, as their 'religion' was valid.

80 THE LANDSCAPES OF SCIENCE AND RELIGION

'Ultimate' has to mean more than simply 'intellectual'. It has to mean that a believer would categorically 'disregard elementary self-interest ... in preference to transgressing its tenets.'[31] Ultimacy was not simply a matter of intellect but of commitment. The judgement drew attention to Krishna Consciousness's history (the movement 'is an outgrowth of the ancient and diverse Hindu tradition'), its geographical spread and popularity ('we think it significant that the Krishna Consciousness Movement has attracted thousands of devotees throughout the world'), and its doctrinal detail ('even if viewed apart from Hinduism, [it] has an "elaborately articulated body of religious doctrine"') to further support its case.

The Court also remarked that it was significant that the Internal Revenue Service (IRS) had accorded non-profit tax-exempt status to the organisation on the basis of its religious nature, and even though the evidence of the IRS may not have the same metaphysical value as religious doctrine, it is nonetheless worth noting the criteria the IRS use to determine whether an organization should be considered to be a 'church' for federal tax purposes. They are as follows:

- Distinct legal existence
- Recognized creed and form of worship
- Definite and distinct ecclesiastical government
- Formal code of doctrine and discipline
- Distinct religious history
- Membership not associated with any other church or denomination
- Organization of ordained ministers
- Ordained ministers selected after completing prescribed courses of study
- Literature of its own
- Established places of worship
- Regular congregations
- Regular religious services
- Sunday schools for the religious instruction of the young
- Schools for the preparation of its members

This is an extensive list and one that is obviously focused primarily on outward and readily identifiable characteristics of a church. Whereas the courts have recognised the inner dimensions of ultimacy, comprehensiveness, and commitment within the definition of religion, the IRS focuses on the formal, external, and surface signs. Both are clearly important.

It is also worth noting not only that the IRS criteria underline the sheer range of different dimensions inherent in religion, but also the fact that none is necessary or sufficient, the Revenue recognising that not all churches will meet all criteria and that therefore it uses 'a combination of these characteristics' to determine an organisation's status. This is the 'family resemblance' approach in as many words.

It is also the conclusion to which some legal commentators have gravitated when faced with the complexities of Supreme and Circuit Court definitions of religion. Discussing the 'misguided search' for a constitutional definition of religion, George C. Freeman III recognised that the pursuit of an 'essence' of religion, which would be both sufficient and necessary, is futile. In its stead, he too reached for Wittgenstein's family resemblances, and although he recognised that courts 'cannot be content merely to identify similarities and dissimilarities', their job being 'to classify', a Wittgensteinian approach can still be useful in as far as it legitimises the idea that there may be no 'uniquely appropriate classification' and that there will be 'a penumbra of debatable cases'.[32]

Freeman pursued this line of thought by drawing on William Alston's *The Philosophy of Language* to educe eight 'relevant features' of religion:

1. A belief in a supreme being
2. A belief in a transcendent reality
3. A moral code
4. A worldview that provides an account of man's role in the universe and around which an individual organizes their life
5. Sacred rituals and holy days
6. Worship and prayer
7. A sacred text or scriptures
8. Membership in a social organization that promotes a religious belief system.

This is a mixed list. Some, like the last four, Freeman admitted, are religious simply because of their relation to the first two. Sever this and they lose their religious character.[33] Central as the first four characteristics are, however, none is essential (let alone sufficient). For example, a 'supreme being' would deny the religiosity of Buddhists, not to mention many spiritually minded modern people. 'Transcendence' risked ruling out those Ancient Greeks who 'believed in gods but thought of them as somehow part of the natural order', or alternatively risked erroneously including those Kantians

82 THE LANDSCAPES OF SCIENCE AND RELIGION

and Platonists who were committed to the existence of the transcendent 'as a world of "things-in-themselves" or as a realm of essences' but did not attach any 'religious' significance to it. When it came to morality, certain seemingly religious mystery cults had no interest whatsoever in morally guiding their adherents, whereas innumerable ethical codes are clearly devoid of religious connotation. And, however much religious people claim a comprehensive worldview, there are various self-confessedly non-religious ideologies that do the same.

None of this denies the relevance or value of the characteristics of religion that Freemen derived from Alston. Rather, it underlines once again the *necessity* of a 'family resemblance' approach to religion and the apparent inevitability of a liminal space in which religiousness is contested, even if that places the law, with its need to categorise, in an invidious position.[34] Ultimately, however hard courts, commissions, and revenue services try, it is impossible to define religion.

The polysemy of religion: anthropological, philosophical, and sociological contributions

The way in which attempts to arrive at a satisfactory legal definition of religion have inexorably gravitated towards ever greater complexity and uncertainty simply shows the law following, albeit at a distance, the direction of travel taken by the philosophy, sociology, and anthropology of religion. At around the same time as the US Supreme Court was defining religion as the relations and obligations of man to his Creator in *Davis v. Beason*, the German sociologist George Simmel was writing how 'no light will ever be cast in the sybillic twilight that ... surrounds the origin and nature of religion as long as we insist on approaching it as a single problem requiring only a single word for its solution'.[35]

That problematic approach was baked into the very category of religion in the first place, as Peter Harrison has shown. If Christianity (and, in particular, Protestant Christianity) was the lens through which expanding European powers encountered cultures across the world, then the Christian religion became the category against which these cultures were compared, and into which they fitted (or didn't). 'It was the construction of the notion of "religion" in the early modern period, itself premised upon a unique understanding of religious identity forged by the early Christians ... that provided the prototypical model of a belief system for which is claimed universal and transcultural significance.'[36]

DEFINING RELIGION 83

The earliest academic anthropologists of religion, such as E.B. Tylor and James Frazer, interpreted the phenomena they studied as if they were (primitive) attempts to answer the existential questions of life, and/or indirectly to control otherwise uncontrollable natural forces. Mythical stories explained the origins of the world. Divine action explained natural phenomena. Ritual was an attempt to control nature. Religions had their origins in fertility cults, which celebrated, and attempted to secure, the eternal, life-giving cycle of death and rebirth. Belief in spirits and souls was a way of explaining the presence of the deceased in dreams. Belief in and prayer to the gods was a way of managing the vicissitudes of life. 'By religion ... I understand a propitiation or conciliation of powers superior to man which are believed to direct and control the course of nature and of human life.'[37]

Such religious beliefs were superseded by the more formalised creeds and rituals of monotheism, but there was no qualitative difference between them. Tylor, a lapsed Quaker, carried with him a Quaker's distrust of ritual, and Frazer rarely missed an opportunity to draw parallels between biblical stories, Christian rituals, and those of 'primitive' people. Contemporary religions may be more elaborate, sophisticated, and organised than earlier ones, but they were recognisable cousins, both hoping to answer humanity's ultimate questions and needs.

This approach to religion developed into what would be called the 'substantive' (or sometimes essentialist) understanding of religion, which located its essence or key characteristics in its content. That content need not be explanatory in a narrowly proto-rationalist sense, as if religion were only a (bad) way of explaining the existence and form of the natural world, as some modern critics sometimes like to claim. One of the most influential early analyses of religion, Rudolf Otto's *The Idea of the Holy*, centred the idea of religion on experience of the 'numinous', a word he coined to capture the 'non-rational' dimension that he thought foundational to religious experience and life. The word 'numinous' derives from the Latin *numen*, meaning 'spirit' and, according to Otto, it should be understood as the sense of '*mysterium tremendum et fascinans*', to which innumerable religious traditions testified. The numinous was the experience of standing before the presence of that which is 'entirely other', an experience that was *sui generis*, and not reducible to or explicable in any other terms.

Otto's book on the idea of the holy was to impress the Protestant theologian Paul Tillich, who read an early draft in 1917 and subsequently wrote how he had a near-spiritual experience himself when reading it. Tillich was himself to prove an important theorist of religion, whose views were quoted in *United States v. Seeger* and in *Malnak v. Yogi*. In particular, his view that

84 THE LANDSCAPES OF SCIENCE AND RELIGION

faith is the state of being 'grasped' by an 'ultimate concern' which qualifies all other humans concerns has proved influential in discussion of religion, despite being about faith rather than religion. So powerful is this grasp that it motivates people beyond mere passive reception to the state of participation. 'If faith is the state of being ultimately concerned, and if every ultimate concern must express itself concretely, the special symbol of the ultimate concern participates in its ultimacy.'[38] Faith demands involvement.

Otto's idea of the holy was also influential on, among many others, the Romanian philosopher Mircea Eliade, whose work, in particular *The Nature of Religion*, built on Otto's concept of the numinous and separates it from the profane. Eliade developed the idea that hierophany, the manifestation of the numinous, or his preferred term 'the sacred', lies at the heart of religious life. As intimated by Tillich, however, religion is more than simply the *experience* of the sacred, as such manifestations not only reveal but also order reality and invite participation in it. In particular, the idea for which Eliade perhaps became best known was that religious ritual and behaviour enables believers to detach themselves from the secular or profane time in which they live and to re-establish and partake in 'sacred time', most famously in the 'eternal return' to a mythical age.[39]

This idea of participation is essential as it underlines the *active* nature of religion that is in danger of being overlooked by the 'substantive' understanding. It points in the direction of a way of understanding religion as 'functional', best associated with the French sociologist Émile Durkheim. Durkheim was dissatisfied by the substantive/explanatory focus of early anthropologists of religion which, he believed, failed to take into consideration what people actually *do* as part of religion, and what religions actually do to and for people. His late book *The Elementary Forms of Religious Life* steered away from an understanding that emphasised God or the supernatural, which were, he believed, relatively recent developments in the history of religion. Instead, he placed emphasis on religion's social function, the manner in which religious activities drew together and helped cohere otherwise disparate groups of people.

This did not mean that the motivation for or focus of religious activities was irrelevant. On the contrary, Durkheim's famous definition of religion—'a unified system of beliefs and practices relative to sacred things, i.e. things set apart and forbidden—beliefs and practices which unite in one single moral community called a Church, all those who adhere to them'— specifically highlighted the *things* to which the beliefs and practices were related, and called them sacred.[40] However, the sacred here was a broader term that was common, and certainly broader than would have been implied by 'God' or the 'supernatural'. The term opened the door to the idea of civil religion, in which the thing held sacred—a flag, a person, a constitution, etc.—was obviously worldly, elevated to the sacred only through the significance people attached to it.

Durkheim's approach has been criticised in numerous ways, including for unduly downplaying the supernatural element in religion, and for consequently being unable to distinguish between religion and society. Nevertheless, the functional approach to religion has proved influential, being adopted and adapted by numerous scholars, including Clifford Geertz.

Geertz was an anthropologist who adopted from Max Weber the idea of the human being 'an animal suspended in webs of significance he himself has spun', those webs comprising a culture that demanded not an 'experimental scien[tific]' approach but 'an interpretive one in search of meaning'.[41] He brought this approach to bear on his understanding of religion, which he famously defined as a 'cultural system', or more specifically '(1) a system of symbols which acts to (2) establish powerful, pervasive and long-lasting moods and motivations in men by (3) formulating conceptions of a general order of existence and (4) clothing these conceptions with such an aura of factuality that (5) the moods and motivations seem uniquely realistic'.[42] Entirely absent here is any mention of belief in supernatural beings or in God or gods, which, as Robert Bellah noted, 'many current definitions take for granted'.[43]

It would be a gross simplification to claim that all philosophical, sociological, and anthropological definitions of religion fell neatly in either substantive or functional categories, or even that these categories were in themselves sufficient or uncontroversial for conceptualising religion. There are other frameworks for analysis, some of which try to circumvent the dichotomy between the transcendent and the immanent that can develop from the distinction between substantive and functional definitions. Nevertheless, in a crowded field often devoid of clear lines of guidance, the delineation of substantive and functional factors to religion is helpful.

These factors are discernible in contemporary definitions of religion that are frequently cited in the literature. To take one (more recent) example that managed, unusually, to bridge the gap between academic and popular discourse, Ninian Smart was a pioneer of religious studies who also attained a popular reach through his book *The World's Religions* and his work as editorial consultant for the BBC programme *The Long Search*. He proposed a multi-dimensional understanding of religion, which sidestepped the need

86 THE LANDSCAPES OF SCIENCE AND RELIGION

to arrive at a single definition. His (eventual) seven dimensions incorporated both substantive and functional categories, and were, in the order they appeared in his book:[44]

- the *practical or ritual dimension*, which incorporates the specific practices, ceremonies, and patterns of behaviour, commonly but not necessarily public, that are more (or less) important to the religion;
- the *experiential or emotional dimension*, which incorporates the powerful emotional responses—awe, wonder, dread, devotion, ecstasy, peace, etc.—that are provoked by experience of the numinous (Smart references Otto here), the mystical, and the sacred;
- the *narrative or mythic dimension*, which incorporates foundational stories, pertaining to the wider cosmos or to the religion itself, that explain reality, preserve the corporate memories that hold people together, and cohere, to a greater or lesser extent, into a comprehensive narrative of the world (or at least relevant parts thereof);
- the *doctrinal or philosophical dimension*, which works through and systematises the commitments of the narrative dimension, as they come into contact with wider reality which often poses questions to them;
- the *ethical or legal dimension*, which sets out the moral rules for both individual and collective and frequently adds the penalties and rewards for dis/obedience;
- the *social or institutional dimension*, which incorporates the formalised structures, offices, and procedures that characterise (especially older and larger) religious communities; and
- the *material dimension*, which refers to the material culture—places, buildings, artefacts, clothing—that are accorded particular, often powerfully symbolic, significance in the narrative and practice of a religion.

This is a helpful, wide-ranging, and admirably clear list, and one in which there is considerable interlock and overlap. None of the dimensions is detached from all the others. Ritual commonly enacts narrative. The material dimension is central to ritual. The doctrinal and philosophical dimension is parasitic on the narrative, although narratives often claim to explain the originals of doctrinal (and ethical) dimensions. The ethical or legal dimension is often embedded in the narrative and elaborated by the doctrinal. The social and institutional must cohere with the philosophical dimension, and ideally with elements of the narrative. And so on and so forth.

The definitions offered by Frazer, Otto, Tillich, Eliade, Durkheim, Geertz, and Smart, and many others not discussed here, help delineate the concept of religion, drawing the various constituent elements and dimensions within the term. However, the range of definitions, elements, and dimensions on show serves primarily to underline quite how artificial, fluid, and complex the category of religion is. As Charles Taylor observed in a lecture on 'The Polysemy of "Religion"', 'anyone who starts a sentence [with the assertion that] "Religion is ..." or "Religion does ...", without qualification, is inevitably false'.[45] Like the courts and the revenue services, the anthropologists, sociologists, and philosophers have been unable to define religion with clarity and finality.

Conclusion: defining religion

This chapter has outlined a number of high-profile and influential attempts to define religion over the last hundred years or so, drawing on various legal, anthropological, philosophical, and sociological contributions. This is a miscellaneous range of sources and it makes for an extensive, variegated, and sometimes confusing definition.

There has been an enormous shift in the Western conception of religion over this time, from something that resembles Christian monotheism to something much more capacious and amorphous. At the same time, while the law has tried to home in on a workably tight definition, the work of anthropologists, sociologists, and religious studies academics has broadened the category, and generated a widening wake in which the courts have eventually followed.

All that acknowledged, the various elements within the definitions we have examined in this chapter do provide the basis for a family resemblance approach. Religion has substantive and functional features. It usually involves belief in, and perhaps worship of, a 'supreme' or 'transcendent' 'being' or 'entity' or 'principle' or 'ultimate reality'. It aspires to being explanatory at this ultimate level, clarifying and illuminating the meaning of human existence, our place in the universe, and our relationship with that infinite or transcendent reality. It can have a powerful 'emotional' or 'existential' effect on the individual, based on an encounter with the 'numinous', 'holy', 'transcendent', 'mysterious', or 'sacred' that evokes 'wonder' or 'ecstasy', and can 'establish powerful, pervasive and long-lasting moods'.

88 THE LANDSCAPES OF SCIENCE AND RELIGION

While that effect is felt by the individual, religion is also commonly a thoroughly 'social' or 'communal' endeavour, 'uniting' and 'binding' an otherwise disparate group of people in their 'ultimate concern'. It frequently has a profound 'moral' and 'educative' dimension, guiding and instructing adherents how to live their lives. Its beliefs and practices are embedded in 'myth', 'narrative', and 'literature', and are made manifest in a richly 'symbolic', 'ritual', 'material', and 'physical' culture. Its community is often formalised into 'membership', belief into 'doctrine', and behaviour into ethical codes. It often involves a degree of systemisation and structure, 'a certain level of cogency, seriousness, [and] coherence', including the formalisation of its beliefs and the institutionalisation of its moral, social, and ritual elements.

All this is true and constitutes a helpful basis for the disambiguation that we will turn to in the next chapter, but the various caveats deployed above— 'usually ... can have ... is commonly ... frequently has ... is often'—should serve as a reminder of the underlying principle of this whole endeavour, namely that the things we term religion will be characterised by some combination of these features, but not by all or indeed any single one of them.

Notes

1. James, William, *The Varieties of Religious Experience* (London, 1994 [1901–02]), p. 31.
2. James, *Varieties*, pp. 36, 31; emphases added.
3. McCauley, Robert, *Why Religion Is Natural and Science Is Not* (Oxford, 2013).
4. *R v. Registrar General, ex parte Segerdal* (1970), para. 709.
5. *Re South Place Ethical Society, Barralet v. AG* (1980), para. 1573.
6. 'Everyone has the right to freedom of thought, conscience and religion', Equality and Human Rights Commission (EHRC), Article 9: Freedom of thought, belief, and religion (www.equalityhumanrights.com).
7. 'Freedom to change ... [and] manifest ... religion or belief, in worship, teaching, practice and observance', EHRC, Article 9.
8. Employment Equality (Religion or Belief) Regulations 2003, SI 2003/1660; emphasis added.
9. Originally section 2(3)(a) of the Charities Act (2006), but section 3(2)(a) of the Charities Act 2011.
10. UK Government, *Analysis of the law underpinning The Advancement of Religion for the Public Benefit*, Section 2.25(a) (2008) (chrome-extension://efaidnbmnnnibpcajpcglclefindmkaj/https://assets.publishing.service.gov.uk/media/5a7d5924e5274a2af0ae3120/lawrel1208.pdf).
11. *Application for Registration of the Gnostic Centre* (16 December 2009), para. 23.
12. UK Government, *Analysis* (2008), 2.25(b); *Registration of the Gnostic Centre* (2009), para. 23.
13. UK Government, *Analysis* (2008), 2.2; given slightly differently as 'cogency, cohesion, seriousness and importance' in *Registration of the Gnostic Centre* (2009).

DEFINING RELIGION 89

14. UK Government, *Analysis* (2008), 2.3–2.4. Or 'an identifiable positive, beneficial, moral or ethical framework' in *Registration of the Gnostic Centre* (2009).
15. *R (on the Application of Hodkin) v Registrar General of Births, Deaths and Marriages* (2013), para. 34.
16. *Hodkin*, para. 51.
17. *Hodkin*, para. 61–62.
18. 'I emphasise that this is intended to be a description and not a definitive formula', *Hodkin*, para. 57.
19. *Hodkin*, para. 57.
20. *Hodkin*, para. 34.
21. Quoted in Freeman, George C., 'The misguided search for the constitutional definition of 'religion', *Georgetown Law Journal* (1983) 71, 1520.
22. *Davis v. Beason* (1890), para. 32.
23. *United States v. Macintosh* (1931), para. 34.
24. *United States v. Ballard* (1944), para. 86–87.
25. *Fowler v. Rhode Island* (1953), para. 69–70.
26. *Torcaso v. Watkins* (1961), para. 495, n. 11.
27. *United States v. Seeger* (1965), para. 165–66.
28. *Welsh v. United States* (1970), para. 342–343.
29. *Malnak v. Yogi* (1979); emphasis added.
30. *Malnak*, para. 207–208.
31. *International Society for Krishna Consciousness, Inc., v. Barber* (1981), para. 440.
32. Freeman, 'Misguided search', p. 1552.
33. Freeman, 'Misguided search', p. 1554; 'A mystic might avoid affiliation with any religious organization; a primitive religion might sustain itself without any literature; and a Buddhist might not worship, pray, celebrate holy days, or practice any rituals.'
34. Alston, William, *The Philosophy of Language* (Englewood Cliffs, NJ, 1964), p. 90. This is entirely in line with Alston's own reasoning: 'Even if we could say exactly which or how many of the various religion-making characteristics a [belief system] has to have in order to be religious, we would be unable to say with respect to a given characteristic, exactly what degree of it we must have in order to apply the term.'
35. Simmel, Georg, *Essays on Religion* (New Haven, 2013) p. 101. As it happened, Simmel hoped (unsuccessfully) to establish such a definition. His analysis of the basic problem was more acute that his proposed solution.
36. Harrison, *Territories*, p. 191.
37. Frazer, James George, *The Golden Bough: A Study in Magic and Religion* (New York, 1925 [1890]), p. 50.
38. Tillich, Paul, *Dynamics of Faith* (New York, 1958), p. 122.
39. Eliade, Mircea, *The Myth of the Eternal Return: Cosmos and History* (Princeton, NJ, 1971).
40. Durkheim, Émile, *The Elementary Forms of Religious Life* (Oxford, 2001 [1915]), p. 67.
41. Geertz, Clifford, *The Interpretation of Cultures* (New York, 1973), p. 5.
42. Geertz, *Interpretation*, p. 90.
43. Bellah, Robert, *Religion in Human Evolution* (Cambridge, MA, 2011), p. xiv.
44. Smart, Ninian, *The World's Religions* (Cambridge, 1998), pp. 10–21.
45. Charles Taylor for IWM Vienna, *YouTube*, 'Charles Taylor: The Polysemy of "Religion"', 17 May 2018.

5

Disaggregating Religion

Introduction

As with science, so with religion: just as experts could and did offer definitions of science, so they could and did for religion. But, as with science, many were reluctant to do so and keen to point out the problems, verging on the impossibility, of the task.[1]

For some, the problem lay in the plurality terms available. It was, according to one, linguistic and conceptual 'chaos out there' (#5). The term 'religion' was endlessly blurred with other terms and ideas with which it was often interchanged, such as 'spirituality', 'belief', 'faith', 'philosophy', 'way of life', and 'worldview'.[2] Religion was 'a very problematical category' (#97), with hardly any 'boundary conditions' (#83). The category was 'not a very well-defined thing' (#13), problematically sprawling—'so hugely over-capacious and embrac[ing] so many different things that any effort at trying to "precisify" it, is very controversial' (#61)—to the point of being incoherent. 'There's a lot of debate about whether it exists, whether it's a concept that even is valid as a coherent whole' (#84).

Some highlighted the irreducibility of religion*s* to religion.[3] Others pointed out the impossibility of defining categories without contexts.[4] A number highlighted the 'Buddhism and football' paradox (cited in chapter 1),[5] or versions thereof, such as Buddhism and Protestantism,[6] Buddhism and Marxism,[7] Buddhism and theism,[8] or football and politics.[9]

Religion was judged an artificial and 'constructed' category, not simply in the sense that all linguistic categories are artificial and constructed—they could hardly be otherwise—but *problematically* constructed, mapping badly onto the reality it purports to signify. For some, this was an accident of history. Those who understood the intellectual hinterland pinned the terminological inadequacy of 'religion' on its recent and circumstantial history.[10] According to one philosopher, the problem was more contemporaneous, as he claimed that definitions were often really a function of social and intellectual pressure.

Even people who are religious are often themselves not terribly clear about what it is that they're doing, and they'll adopt different understandings of religion depending on whom they're talking to and how convenient it is to them. (#42)

Some interviewees (far more so than in comparable discussions about the definition of science) concluded that the category was little more than a fiction, with little correspondence to recognisable reality. 'It is something that humans think up, rather than something that extrinsically exists in the world'. (#7) A few simply gave up on any attempt—'I'm not even sure where to start in that ... I am going to opt out of this one if you don't mind' (#57)—but most persisted, gravitating, wittingly or otherwise, to the two Wittgensteinian approaches highlighted in the Introduction.

First, echoing Wittgenstein's admonition 'Don't think, but look!', several claimed that 'You'll know it when you see it' (#58). Religion could be identified by observing what people did or, failing that, listening to how they talked about it. Second, recognising, in Wittgenstein's terms, that religion was 'a complicated network of similarities overlapping and criss-crossing', some claimed that it was 'not always one thing but is perhaps a package of several different things in human thinking and behaviour' (#66), while others concluded directly that 'perhaps there's something more like family resemblances going on' (#101).

Adopting this 'family resemblance' approach in the same way we did with science, we can draw out five features with the family resemblance of religion:

- Transcendence
- Belief
- Commitment
- Formalisation of belief in theology and doctrine
- Formalisation of commitment in rituals, laws, and offices

These are described in greater detail below, although, as with science, the order does not imply any hierarchy.

Transcendence

The question of 'transcendence' was, along with that of community discussed below, the most commonly mentioned element within discussions

92 THE LANDSCAPES OF SCIENCE AND RELIGION

of religion. The word, however, merits scare quotes because the vocabulary used here was conspicuously broad.

A minority spoke simply of God, although even when they did so, the term was often complexified: 'the existence of God, or whatever that means' (#26). More amorphous terms—like 'deity' (#38), 'entity' (#12), 'supernatural agency' (#61), or 'supernatural being' (#21)—were usually preferred. Even with these, however, it was commonly noted that the emphasis on a higher *being* was a particular ('Western') view that was not necessarily replicated elsewhere.[11] As a result, a variety of other non-anthropic terms were used, partly to escape the restrictions imposed by Western monotheism and partly as a wider reflection of the linguistic limitations in any discussion of the transcendent.

In this way, the transcendent dimension to religion involved 'belief' (we will come to the meaning of this significant term in the next section) in the existence of 'supernatural entities, usually not visible beings' (#67) or belief in 'one god or many' (#80). Alternatively, moving away from anthropic language altogether, religion meant belief in 'a supernatural element' (#78), 'an unseen world of some kind' (#82), 'a higher power' (#98), 'entities or forces greater than the individual human beings' (#7), 'imagined realities' (#63), or 'an underlying reality and it can be a reality without God, it can be nature, it can be a mystic experience of this' (#90). Ultimately, this variety of terms naturally gravitated to an abstract spiritual language of 'transcendence'[12] or simply to the idea of negation, arriving at an idea of what the divine is by stating first what it is not.[13]

At the same time as there was a general inclination towards the abstract, however, there was a smaller but discernible view which emphasised that, properly speaking, religion entailed a belief in not only the *transcendent* but also a transcendent that was active and personal and interested and involved in the immanent world, and in the world of humans in particular. The nature of this involvement varied. This 'supernatural being or power' had a role of creation, having 'created everything' and being 'responsible for existence' (#36). 'It' had a role of superintendence, 'dominion over the natural world' (#35), 'look[ing] over everything' (#36). It had a being and a role that was worthy of praise and honour, 'to whom we ought to be grateful' (#42). Or it had the specific role of 'having powers of moral judgement' (#21). It was 'in some way interested in us, in human beings' (#61) and capable of (some form of) interaction or relationship with us—a point that was made by both one interviewee who was openly Christian and another who was dismissive of all religious beliefs.[14]

If the nature of the transcendent dimension of religion elicited varied reactions, so did the question of its significance and centrality. Was a transcendent dimension essential to religion? Was it, that elusive thing, a necessary (even if not a sufficient) element within religion?

For some, the answer was clearly yes. 'It does seem to me that some connection with the supernatural, to put it again sort of crassly, is pretty much essential to religion' (#28).[15] For others, the metaphysical dimension was largely irrelevant. 'The idea that God created the universe, whatever you understand by that, I guess I don't really understand why that should be an essential component of religious belief' (#10).

The division turned, in some measure, on the same normative/descriptive tension that ran through the definition of science. No matter what religion was in a normative sense—what religion was when it was being real religion—some were insistent that, descriptively speaking, many had an understanding of religion that was largely indifferent to the supernatural.

> Some people understand religion in a very metaphysically non-committal way. They would think of religion in terms of communities, traditions, practices and not in terms of claims as to the nature of reality and so religion understood in that way, well that's one way of understanding religion. (#55)[16]

The implicit question was, was such a religion *real* religion? Others were equally insistent that, whether or not there were people who were primarily, even exclusively, interested in the purely functional, communal concept of religion, religion *normatively speaking* had to integrate some belief in the transcendent. The judgement could not be settled by observation alone because one party could point to examples of functional-only religion, whereas the other could point to the fact that there were (far) more examples of religious practice with a clear, explicit, and non-negotiable commitment to the transcendent.

In this regard, the issue of transcendence not only assumed central significance to the question of what religion was, but also underlined the necessity of a family resemblance approach. Certainly not sufficient, and probably not even strictly speaking necessary, the dimension of transcendence was nonetheless one of the two 'family features' that most people looked for when trying to identify whether something was a religion. The other, a communal dimension, we will turn to later.

94 THE LANDSCAPES OF SCIENCE AND RELIGION

Belief

The idea of transcendence appears, at first, to be indivisible from the idea of belief. Indeed, it is hard even to talk about the former without doing so in terms of the latter, and people almost invariably yoked the two together: 'a belief in a deity ... belief in some supernatural entity ... belief in some kind of higher power ... belief in some kind of supernatural agent', etc. (#12, #25, #36, #42).

That recognised, the nature and foundations of *belief* are significant in their own right, and merit separate attention. As we shall come to see in part II, there are various perceived tensions in the relationship between science and religion, and those relating to metaphysical commitments—what people believe—are distinct from those relating to methodology—the ways in which they come to, and the reasons they have for, that belief.

What belief (or, interchangeably, faith) meant to people differed, in some measure, according to what was being believed. As one philosopher observed, 'belief commitments ... [are] hugely diverse, ranging from positions I would describe as deeply irrational to ones that are highly intelligent and highly thought through' (#5). In the light of such a distinction, belief could be more or less reasonable, and more or less well evidenced.

For a good number of people, it was definitely less reasonable and less well evidenced. Indeed, in a way that had significant implications for the perceived relationship between science and religion, belief was repeatedly defined as something that happened either in the absence or in the teeth of evidence.

> We take faith to mean a belief of things regardless of the evidence for them ... the key things for faith that I would say [are] it is valuing belief despite evidence, or even because of the lack of evidence ... one of the things about faith is it's supposed to stand up whether or not there's evidence. (#1, #9, #19)

This was (self-evidently) the view of interviewees who were non-believers. Other non-believers demurred and insisted that there was some evidential basis for belief, while believers (naturally) had a more positive basis to the evidential and rational basis of their beliefs.

The underlying question was what constituted evidence. Even those most antipathic to religious belief recognised that there was a process of reasoning and argumentation within religious believers' approach. The problem, as far as they were concerned, was not so much that the reasoning was bad (although that could be a problem) as that the evidential basis was illegitimate. That which some considered to be evidence was ruled as mistaken, inadequate, or inadmissible by others.

In contrast to this, many people—and not only believers—insisted that such 'alternative' sources of evidence, to which we shall turn presently, were perfectly legitimate. Indeed, they were necessary and inevitable. After all, the argument ran, if religion is a holistic (rather than narrowly cognitive) phenomenon, which integrates and demands a whole life response, it is only natural and to be expected that there be a range of evidence to include the existential and personal.

At a generic level, people spoke of 'feelings' or 'convictions' as contestably legitimate sources of evidence for belief. 'Faith starts with a feeling of belief that you have, not with some kind of concrete' (#48).[17] More precisely, there was the widely attested experience of awe and wonder. 'Religion to me is actually a normal human experience. We are constantly struck by overpowering sensations of wonder, of mystification, of love of all things that religion tells us stories about' (#54).[18] There was the experience of connectedness to and within the universe, not just being awed by creation but feeling profoundly and meaningfully linked to it, 'a sense of being part of the cosmic universe ... this feeling of personalised immersion in some greater spirit' (#51). There were experiences pertaining to the coherence and order of the universe, 'a sort of intuitive and slightly vague set of just feelings about how the world operates and what might be going on' (#13). And there were personal experiences themselves, individual and unique rather than generic and universal, moments of individual illumination, sometimes when undergoing 'religious' activities, but not necessarily so. 'I think for most people, it's not so much an articulate belief as a matter of experience, that fact that they can pray and they feel there's some consciousness to which they are praying' (#41). People believed, or were certainly thought to believe, for a variety of reasons, but high among them were a sense of awe, wonder, connectedness, order, or the personal experiences that affected or even transformed them.

This perspective on belief offers an answer to the enduring questions of how far the 'belief' of religion pertains to the physical world, and to what extent is belief considered explanatory. There were certainly some who thought the answer to both questions was affirmative, at least in origin. 'I personally tend to think of [religion] as early attempts to explain the world, an incomprehensible world' (#19). Here, we found evidence of what is

96 THE LANDSCAPES OF SCIENCE AND RELIGION

sometimes known as the theory of HADD, a hyperactive (or hypersensitive) agency detection device.

> Our cavemen ancestors, when they at length had the leisure to meditate a bit on the world around them, to try to come up with some sort of explanation for natural phenomenon, the one tool they had was their own experience of agency … Some agent is making it happen and indeed when we inspect mythologies, we see that every stream has its nymph and every tree its dryad and the earthquake is caused by Poseidon and lightning is thrown by Zeus and this is a big elaborate explanation in terms of agency which has been anthropomorphised, given names and given traditions. (#61)

Religion, on this reading, explained the origin, form, and processes of the natural world—or, at least, that's how it originated. For a few, that was still its basis today. Religion 'probably comes from the same impulse that science comes from as in a desire to understand more deeply what's going on around us but from a very different philosophical perspective' (#13). At its crudest, it 'invokes a supernatural element to explain phenomena' (#78) or 'cites supernatural intervention as an explanation of events' (#46). At a more sophisticated level, 'a belief in a deity … explains [the world's] intelligibility, its structure, the fact that it is organised' (#12).

This was, however, a minority view. Far more common was the idea that the belief inherent in religion was oriented to human existence rather than that of the natural world; that, in essence, religion was about understanding the human condition rather than explaining the natural one. Time and time again, interviewees stated how the significance of religion lay in 'exploring our existence and our place in the world' (#71), or 'the desire to understand the meaning of your existence in the context of the meaning of the whole of reality' (#17), or the 'search for something transcendental in order to try and make sense of mysteries' (#20), or in 'trying to explain human existence and the need for human existence' (#99). It was this view that predominated, irrespective of belief or area of expertise. It came from sociologists of religion:

> Religion is also about the meaning and purpose and value questions, so it's about why are we here? What's the purpose of existence? Where do I find meaning in my life, and then what should I do with my life? What

are the good things to do? What are the not so good things? What are my obligations if I have them? (#28)

From cosmologists:

> The only interesting philosophical question, or at least the most interesting one, which is what does it mean to live a finite, fragile life in an infinite and possibly eternal universe ... It's certainly very big and very old, and we are certainly very small, and very fragile. So, I think that's an interesting question and I think to my mind religion and most of philosophy and naturally my personal thoughts are associated with understanding that question. (#77)

From philosophers:

> Religion is something that is essentially tied up with what it means to be human and questions of meaning and morality and redemption and love, all of these categories are not scientific categories and to try and squeeze religion into that straight jacket is really distorting. (#93)

From theologians:

> [Religion is] a human art devised by the human animal to try and explain and understand its own mysterious existence in a mysterious universe. (#73)

And from anthropologists:

> Nearly everybody who is human has those big questions. Who are we? What are we doing? Why are we here? And if you seek explanation for that, it is not necessarily going to be found through material things, and not necessarily within scientific explanation ... even if it doesn't completely answer these big questions, it provides a response to them. (#66)

In as far as there was a single, necessary element within the definition of religion, this was it: the attempt to formulate an answer to the existential questions that are inherent to human life. In reality, it wasn't technically *necessary*, as there were certainly some conceptions of religion that placed singular emphasis on its functional/communal dimension, at the expense of even this relatively diluted substantive dimension.

98 THE LANDSCAPES OF SCIENCE AND RELIGION

Moreover, however commonly mentioned, it was not sufficient, as everyone agreed there was more to religion than answering such existential questions, and neither was it distinctive, as plenty of people pointed out that self-consciously non-religious belief systems, worldviews, and philosophies also attempted to answer these existential questions.

Commitment

Belief, when used of religion, could refer to the intellectual assent to the existence, presence, relevance, or interference of some non-material, transcendent person, being, or dimension—but it was not limited to that. Belief also meant 'placing trust' or 'having confidence' in that person, being, or dimension. Belief was affective and existential as well as cognitive, and by dint of being affective and existential it required, and was characterised by, a holistic response rather than simply intellectual agreement. In effect, religion also had an important dimension of personal, existential commitment. 'You could define religion as ... committing you to some more powerful being' (#55).

For some people at least, this was arguably more important than the dimension of belief as intellectual assent. 'I don't think propositions are particularly important. I think it's our relationship with other beings that is religion' (#67). Religion was not simply about believing in things but 'finding a belief system [we will turn to the systemic nature of belief below] that will see you through the trials and tribulations of life' (#37). It was belief with a purpose, an attempt 'to satisfy or assuage a general fear of existence, or a struggle to try and place oneself in the world' (#37). Religion was an 'existential culture that involves a belief in God' (#30).

For a few people, this constituted a relatively straightforward either/or, with the commitment of religion being self-evidently more important than the cognitive beliefs.

Many people misunderstand your question I think, by assuming that religion is a series of statements, some of which are valid and some of which are incompatible with observation. But I don't wish to define it as that, because then it can be easily dismissed as a bunch of nonsense ... it's a community, it's a set of behaviours, it's not necessarily a coherent belief system. (#77, #13)

More, however, recognised that existential commitment did not necessarily mean that cognitive belief was irrelevant or even less important. It was, rather, that such belief, when not accompanied by an existential element, was deracinated or partial, a shadow of what it should be. 'I suppose you could have a very strong sense of beliefs, but not act on them in any way, shape, or form, and they have no effect on your life whatsoever. But then I guess I would question whether you really believe them' (#53). Religious commitment was the fruit of the plant whose roots were religious belief.

What, then, to press the metaphor, was this fruit? At a generic level it was 'a set of practices or behaviours or ways of living your life' (#91). More specifically, the answers to this question clustered loosely around three points. The first was morality. Religious commitment integrated an ethical dimension. This could be narrowly conceived as the business of following God's strict and restrictive rules. 'Religion is always about saying, "In this vast universe of things you could do, only these are good. God only permits these things. You must do these things"' (#21). Alternatively, it could be seen more broadly as reflecting on the ethical implications of belief in the transcendent. 'It seems much more significant to ask ... what that understanding [of God] says about the way we should live our lives' (#10).

Crucially, this ethical dimension involved more than simply a set of moral beliefs in the narrow sense of believing that x was right or y was wrong. Rather than merely being the assent to certain values, it involved doing or practising moral commitments. Religion is 'a way of being in the world, by that, I mean physically, bodily, materially, socially' (#63). The material element was crucial. In the words of another, 'it's a bodily thing ... an incredibly bodily thing. People fast for days and weeks. It is very much about your physical body and your physical presence on earth' (#38).

The example of fasting links to the second element, namely that religion involves a degree of ritual commitment. For 'many people, religion is a matter of ritual and custom ... it's something which provides guides for life and elaborate common rituals' (#33). The nature of these rituals varied enormously, ranging from the formal, precise, and legalistic to the casual, occasional, and non-committal. Moreover, as some acknowledged, a focus on ritual risked generating a rather circumscribed notion of what religion is. 'Religion in its narrow sense [is] the execution of various rituals and ceremonies and so on' (#45). Furthermore, as already mentioned, some believing respondents distanced themselves from the idea of ritual altogether in the same way as they did from the idea of religion. They were uncomfortable describing their faith as a religion, let alone a ritualised one.

100 THE LANDSCAPES OF SCIENCE AND RELIGION

Nevertheless, these points acknowledged, ritual was widely judged to be significant, albeit for unclear reasons. The perceived purpose of religious ritual or ceremonies varied. In origin, it was speculated, they might have been the counterpart to the explanatory dimension of religious belief. If religious beliefs once helped explain the way the material world was, religious rituals helped manipulate it.

That was not the view of religion today, however, where it was more commonly assumed that just as religious beliefs helped locate humans' 'mysterious existence in a mysterious universe', so religious ritual commitments were a way of enacting and navigating that existential location. As one interviewee put it, 'religion is a set of bodily behaviours and practices that index imagined realities ... imagined beings' (#63).

This helped explain religion's particular ritual interest in seminal moments of human life.

> It provides rites of passage, births, marriages, deaths that give significance and meaning to life. It provides liturgical year, a cycle of festivals and holy days that structures life and gives meaning individually and communally ... it provides a series of practices like prayer, which enable people to connect and feel more connected with realm of consciousness beyond themselves. (#42)

Ritual was a way of locating, marking, understanding, and therefore equipping and sustaining humans within the cosmos—material, moral, existential—in which they found themselves.

Both the ethical orientation and ritual practice within religious commitment could, in theory, be individualised. Private prayer and personal contemplative practices were cited as examples of this. However, without a doubt, the dimension of religious commitment that was most frequently highlighted and judged as most significant was its community, the third element of religious commitment.

This point was made etymologically by several interviewees. '*Religio*, the kind of root of that word, is binding together and so for me what is distinct about religion is, it's a communal thing' (#74).[19] The etymological fallacy notwithstanding, this was a very widely held view of religion. Ethical beliefs, ceremonies, ritual, and worship were primarily communal affairs. Religion should 'be viewed as a social institution' (#10). 'It has to have some collective element in it' (#80). It was 'characterised by communities that form around particular rituals and particular commitments about ultimate reality, which result in certain types of life practices' (#100). 'The true value of religion in all capacities [is] as being community' (#35).

Given how central this view is within the conception of religion, it is striking how absent it usually is in discussions of the relationship between science and religion. And yet, as we shall see in part II, it is a highly germane issue.

Formalisation of belief in theology and doctrine

As mentioned earlier, one of the reasons why it is difficult to define religion is the plethora of seemingly interchangeable words that were used to describe the phenomenon. The way in which people most often attempted to draw a distinction here, at least with the most popular cognates—say between religion and spirituality, or religion and faith—was through the notion of formalisation. 'Religion ... has the connotations of [being] formalized ... and there being a structure as opposed to personal belief or faith' (#9).

Formalisation worked on two separate dimensions of religion—belief and commitment. The former is explored here, the latter in the following section. The first thing to acknowledge is that there were some interviewees who were adamant that such formalisation was anathema to real religion. In the words of one psychologist:

My whole pitch here is that religion has nothing to do with theology. [Religion is] something to do with the raw feelings as the philosophers used to call them ... my sense is it's very much about real relationships as they're experienced ... the kind of experience you have with the relationship with God, whoever it may be. (#51)

This, however, was a minority view, and the understanding of religion on show here—'the kind of experience you have with the relationship with God'—is closer to what more people termed spirituality. More people recognised that formalisation was, for better or worse, an intrinsic element of religion.

Precisely what formalisation entailed was unclear—a point that was indicated again by the variety of terms used to describe it. Religion was faith or experience that was 'organised' (#29), or 'codified' (#92), or 'reinforced' (#95), or, most commonly, 'systematised'.

102   THE LANDSCAPES OF SCIENCE AND RELIGION

> Religion is the formalisation of certain aspects of faith. If you like, religion is when you end up developing a belief system, particularly a belief system in something beyond ourselves. (#70)

Moreover, as we shall note in the following section, the formalisation of belief into doctrine or theology was inseparable from the formalisation of commitment into ritual, laws, and offices.

There were different opinions on how desirable this was. At one end was the view that formalisation was simply a necessary and inevitable counterpart to religion's communal nature. If religion involves belief *and* communal commitment, it is inevitable that it will try to organise and codify those beliefs so that differences don't end up undermining that communal commitment. (Of course, as interviewees knew only too well, such attempts to prevent differences in belief from undermining communal commitment often ended up catalysing the very thing they set out to avoid.) Nevertheless, the point here was that the formalisation of belief was a neutral and arguably necessary process.

Rather more common was the view that such formalisation was somehow a betrayal of, or at least a fall from, what religion could be. In the former category, one interviewee lamented that 'the problem with faith traditions is that they want to scientise themselves. They are not content to just be faith traditions' (#73). According to this view, precision, organisation, and systematisation were the enemy of real religion. Apropos the latter, another interviewee remarked on the (sad) difference between spirituality (a good thing) and religion (a bad one).

> In public life ... spirituality is taken to be a good thing. That is an openness to the beauties of nature and openness to seeing beyond this world that we live in, being non-materialistic ... religiosity, on the other hand ... is basically blindly following the dictates of a very meticulous and sometimes overweening deity. (#89)

These quotations, and the broader discussion around the formalisation of religious belief (and, even more so, commitment), point towards a critical dimension within the science and religion debate, namely the accretion of morally loaded connotations to each category. Science, for better or worse, is often (though by no means always) used in a positive register; religion is often (but not always) used in a negative one. This is for different reasons usually connected with media and public discourse, as we shall see in the next chapter, but in the case of religion as formalised belief and commitment, the negativity is heavily influenced by the individualised, sceptical, anti-hierarchical, and anti-institutional culture dominant in the West. To return to the etymological roots, the idea of binding people together is not popular today, in a culture in which association and community must be voluntary, contractual, and revisable in order to be legitimate. Whereas the communal nature of religion was approvingly cited by numerous people, including those without any religious belief or commitment themselves, the theoretically binding nature of the community was much more problematic.

Formalisation of commitment in rituals, laws, and offices

Just as religion involves the formalisation of belief, so it also involves the formalisation of commitment, or, as one science communicator put it, the 'institutionalisation of ... practice' (#80).

This dimension imitates the pattern of the previous one; indeed, it draws upon it. The formalisation of commitment takes its cue from the doctrine that emerges from the formalisation of belief. Or, put more simply, the rules and rituals and offices that are part of religion derive their legitimacy from the underlying system of belief.

> Once you have got this notion of—not just authoritative—but a form of divine writing, scripture, that is not to be challenged, then the cerebral aspect, and the intellectual notion of religion formally ... in terms of religious authorities, clerics and priests ... formally eclipses behavioural, bodily, social practices. (#63)

As faith is systematised into doctrine, so commitment is systematised into laws, ethical codes, codes of practice, observances, offices, hierarchies, structures, rituals, ceremonies, and the like (#11, #12). If 'faith is living ... as if certain things were true and mattered and have an impact', then 'religion is the structures that have grown up to support that living' (#96).

The extent of those structures varied between (and within) religions, and, in some instances, 'religion has become a very individualised [phenomenon]' (#96), with fewer formalised rituals, limited hierarchies, etc. Moreover, as with formalised belief, the extent to which the formalisation of

104   THE LANDSCAPES OF SCIENCE AND RELIGION

commitment was a positive (or at least neutral) or a negative phenomenon also varied.

As the quotation above suggests, some could see religious structure as a means of 'support[ing]' living faith. In a similar vein, religion was described as 'a combination of practices that are agreed upon by a specific community that then are followed through in traditions and practices and include both those practices and beliefs' (#82). Such formalisation was intrinsic to what religion was rather than some alien or corrupting accretion. Religion is 'any formal organisation that is held together by a series of narratives and perspectives and rituals which include the transcendent as a central dimension to the way in which the knowledge is understood and worked out' (#34). The formality and organisation is part of the package.

Conversely, the formalised commitment could be seen as a real problem. By one account, it was parasitic on formalised belief. Even if the systematisation of belief was inevitable for religion, the systematisation of the commitment was not.

> I'm not sure that institution is the right word. But it is a collective group and again with a hierarchy in it ... religion for me is that subject, that doctrine, whereas added in to the religion is, of course, the hierarchy of the structure within that. (#69)

The result of this was that there could be a degree of flexibility around, for example, religious hierarchy or ritual, while there was less around, say, doctrine or ethics.

More critical still of the formalisation of commitment were those who saw it as essentially antipathetic to any potentially illuminating, liberating, or empowering capacity of faith or spirituality. Formalised commitment that came at the expense of personal belief in or experience of the transcendent risked constricting people. The view was commonly expressed by juxtaposing religion with faith or spirituality.

> Without faith all you have is a fairly hard and uncaring structure that might well give you rules that you kneel down here, you stand up here, you hold this here, you do whatever, it becomes a ritual and a religion rather than a faith. A faith then is something for me that is personal, it is a belief. (#69)[20]

In the same vein, but more pugnaciously still, if that personal element does die, all we are left with are institutions for the exercise of raw power.

Religion serves the interest of the religious class. There is a class that benefit from religious protocols. There is an establishment. There is a hierarchy. It's a way to exercise power in society. (#71)

Conclusion: outliers and dissensions

As with our disambiguation of science, the purpose of this chapter has been to listen to the ways in which people talked about religion and to draw from that exercise a series of features—in this case five—which comprise the 'family definition' of religion. Between them, the ideas of religious transcendence, belief, commitment, formalised belief, and formalised commitment capture the features that characterise the family of religion.

This does not mean that these dimensions fully summarise all the discussion of religion. As with science, there were understandings of religion that were seemingly distant from any of these elements. Thus one interviewee, an archaeologist, outed themselves a 'Durkheimist', with their conviction that 'to be social is also to be religious'.[21] A second spoke like a Freudian, although without owning the label: religion is 'the projection of desire ... [one of] the deepest desires of humankind ... [is to] have a kind of father figure who will protect them'.[22] A third defined religion as 'a patriarchal system of power that is used to subjugate women' (#71), a fourth described religions as 'wicked memeplexes that grasp onto all these very natural things', (#24), while a fifth said, rather less hostilely, 'religion is those practices that can help you being saved' (#90). It is important not to lose sight of these outlier definitions altogether as we draw together the family resemblance of religion.

The family resemblance approach adopted in this analysis emphasises that an activity need not encompass every one of these features to warrant being called 'religion'. It also stresses that other activities, not usually labelled 'religion', will exhibit some of these 'religious' features. After all, personal spiritual practices often integrate the transcendent, science requires belief,[23] campaigning organisations demand commitment, ideological groups recognise codified systems of belief, and plenty of institutions—from universities to courts to sports tournaments—have formalised practices. To repeat, the dimensions outlined above are not intended to constitute necessary, sufficient, or boundary conditions.

This means that, as with science, there can be no definitive in-or-out element to this definition, and there was naturally disagreement on the relevance and importance of the different features outlined. We have already

106   THE LANDSCAPES OF SCIENCE AND RELIGION

indicated this in passing, above, but it is worth emphasising here, not least as it speaks directly to the distinction, cited in the previous chapter, concerning substantive and functional understandings of religion.

A sense of the transcendent was a significant dimension of religion—except for those people who thought otherwise.

> Religion has to be viewed as a social institution ... The idea that God created the universe, whatever you understand by that, I guess I don't really understand why that should be an essential component of religious belief. (#10)

The transcendent was important for religion—but not exclusive to religion.

> You can have all sorts of supernatural beliefs that haven't got anything very much to do with religion. (#30)

Beliefs are important—except among those who see them as thoroughly subordinate to practice.

> The way that analytic philosophers think about religion, is just ridiculous. This idea that what's going on with Christianity, say, is that you sign up to a bunch of propositions and then if asked you can justify them, that's just absurd. It's all about community and ritual and belonging and all of that stuff. (#64)[24]

Formalised practice, such as ritual, was important—but hardly essential or unique:

> I don't think you can define religion in terms of ritual. Secular people have rituals. They're an element of religion, some religions, most religions, all religions perhaps. (#67)

Formalised belief is important—but not for all movements that are labelled 'religion':

> Western notions of religion are very codified and distinct. It feels like a tangible thing. Whereas I think religious traditions I grew up with we didn't even think of them as religion. (#48)[25]

Some religions have formalised commitment—but only a questionable sense of the transcendent and none of the divine.

> Some religions have rules, but they have no gods. Like certain forms of Buddhism and so on. (#94)

Religion encompassed all these different dimensions, but to different degrees and in different ways. 'A propositional approach to the divine realm is not central and I wouldn't want to dismiss altogether, but I think to make it central would be in my view an error' (#75). Attempts to articulate a generalised dimension that was necessary and sufficient tended to dissolve the category altogether. 'Maybe the thing that all religions have in common is that they're somehow or other interested in the meaning or purpose of life, the basis of morality, that kind of thing' (#64).

The result is religion as a spectrum, 'ranging from a belief in some kind of supernatural agent or agents to whom we ought to be grateful ... moving along that scale towards the ... ineffable ... [and then to] more about a kind of ethical commitment and a social belonging' (#42), or, as we have been discussing, a family with various possible resemblances, some of which, as we shall see, are felt (by some) to be in tension with (some of the family features of) science, and some of which that aren't.

Coda: conflict by definition

Carefully defining—or disambiguating—the categories of science and of religion is essential to the task of pinpointing where precisely the alleged tensions between the two reside. The conclusion of part I will gesture in this direction, and part II will explore some of the various different flash points in greater detail.

However, this process of disambiguation also draws attention to another important source of tension between science and religion, namely *conflict by definition*. This refers to the way in which science and religion are judged to be in conflict (or tension or contradiction, etc.) with one another simply because they are defined that way.

This was discernible in many of the discussions around the definitions of science and religion, although it took two apparently opposed forms. The first was, straightforwardly, the idea that science and religion were defined against one another. Science was that which was not religion. Religion was

108 THE LANDSCAPES OF SCIENCE AND RELIGION

that which was not science. That *was* their definition. Sometimes this was every bit as blunt as that summary. Religion is 'a set of rules and a way of understanding the world in a non-scientific fashion' (#39). Religion is 'a way of thinking about the world that is not scientific' (#49). More often, this oppositional definition was visible through the methodological or epistemological lens. For example, 'everything that I can falsify is science. Everything that I can't falsify is religion' (#27), or 'one is a belief system, and one is a disbelief system' (#37), or 'experiment and empiricism and relating measurements to the theories of the real world [this is] science [whereas] religion relies on faith' (#50).[26] By this reckoning, primarily on account of its approach to knowledge, religion was, by definition, the opposite of science.

The second basis of conflict by definition was, paradoxically, the opposite. Science and religion were not the opposite of one another but, in effect, the same thing or, more precisely, grounded in the attempt to do the same thing. Superficially, this spoke of harmony, and it could certainly be used to derive a relationship of complementarity.

> I think [religion] probably comes from the same impulse that science comes from as in a desire to understand more deeply what's going on around us but from a very different philosophical perspective ... Both science and religion as essentially being truth-seeking enterprises ... I think of scientists and theologians as using different tools to get to the same reality. (#13, #23)

However, this similarity between science and religion naturally risked slipping into competitiveness, either because science proved to be an inadequate attempt to answer existential questions or, more commonly, because religion proved to be an inadequate attempt to answer material questions.

This could feed on the view of religion as (an early and now-discredited) explanatory endeavour.[27] In as far as science and religion were both attempts to explain material reality, they inevitably ended in conflict as science bested religion's explanations. But you didn't need to adopt that understanding of religion's function to see tension here. Even if science and religion were mere attempts to understand the same thing—sociality, consciousness, morality, etc.—ultimately it was likely that one would be shown to be a better way of understanding than the other.

In these two ways—either by defining one entity as the opposite of the other or by defining them as similar but therefore ultimately competitive enterprises—science and religion were automatically placed in incompatible positions. The tension between the two was made self-fulfilling, the relationship between them *by definition* conflictual.

By no means did everyone define them this way. Indeed, this 'conflict by definition' was a minority view, with which some disagreed vigorously.[28] But its very existence serves as another reason to wrestle carefully with the meaning of science and of religion, before we plunge into discussions of the relationship between the two.

Notes

1. As we shall note in the relevant section below, there were usually exceptions to most views expressed, and this point is no exception. Thus, one scholar said, 'I think what a religion is, is pretty straightforward' (#21) and a few others gave short, direct, and unqualified answers, e.g. 'religion is I suppose a faith based organizational principle. That's it' (#87). These were minority views, however.
2. 'It's a mistake to regard Buddhism, Jainism, Confucianism as religions. They're philosophies.' (#61) 'In Islam for example, we don't say that Islam is a religion. We say it is a way of life.' (#59)
3. 'It seems to me there are many religions, and to try and reduce them to some common characteristics is a fool's game.' (#12)
4. 'I hate these working definitions. It is so context-specific, and it depends on which religion you are thinking about.' (#82)
5. 'If we consider football, then fanatical football fans may well be described as religious in some sense, but it's got nothing to do with God or the supernatural, and some forms of Buddhism we easily think of as religious but they also don't necessarily believe in God or indeed anything supernatural.' (#83)
6. 'There is not all that much basically in common between Buddhism and—let's say—Protestantism.' (#12)
7. 'Why is Buddhism a religion and Marxism isn't?' (#64)
8. 'Maybe a religion is people worshiping a god, or gods. Okay. But then there are things you would want to call religion that don't do that in quite such obvious ways. Classic Buddhism being the case there.' (#88)
9. 'Is British football a religion and if you think it isn't, what do you bring into play if you think it's not a religion? ... Is Trumpism a religion? Well, it's a kind of worshipful thing. It's a worship of your team.' (#84)
10. '[Religion] only emerged roughly around the same time as science as a definition, in the nineteenth century. Before [then], religiosity was piety and so on.' (#47) 'Religion as a term has essentially emerged from western experiences of predominantly Christianity.' (#57) 'Aquinas considered religion to be a virtue akin to justice; we now consider religion to be a group of activities normally, or a belief system, pertaining to the supernatural or to a spirituality.' (#83)
11. 'In the West, we tend to think of it theistically, even though ... there are traditions which are clearly religious, which ... you don't have a separation of god and nature.' (#14) It is telling that this interviewee considered these other traditions 'clearly religious', despite their non-theistic nature.
12. Terms commonly used were 'something that goes beyond emotions to transcendence' (#79), 'a search for something transcendental' (#20), and 'practices ... that ... refer in some sense to something beyond, you might call it transcendent' (#84), etc.

110 THE LANDSCAPES OF SCIENCE AND RELIGION

13. This is sometimes known in theology as the apophatic approach, although no interviewees adopted this in a recognisable way.

14. 'As a Christian, I would say that I am not religious because it is actually about relationship, and it is about God revealing himself to me.' (#60) In a similar but also very different vein: 'when you think of the world's young religions, Christianity and Islam, they have adopted many elements of the secret mystery observances like Orphism and so on which had quite a big magical aspect to them and they bought that more intimate and personal thing, an idea of a personal relationship with the deity and a secret inward private aspect to it' (#61).

15. In a similar vein, 'most people, by religion, would mean a set of beliefs and practices predicated on the existence of a supernatural being' (#21) and 'you can have obviously participation and rituals. They're also very important to measure the religion, but unless you have the supernatural beliefs attached to it in some way, for me, that makes it no longer religion the way we would understand it' (#98).

16. In a similar vein, 'for most people, or many people, religion is a matter of ritual and custom, not a matter of dogma at all' (#33) and 'religion tends to be characterised by communities that form around particular rituals and particular commitments about ultimate reality, which result in certain types of life practices' (#100).

17. In a similar vein, 'you don't have clear facts, but you have nonetheless a feeling, a subjective conviction that something is the case' (#45).

18. In a similar vein, religion involved 'a global sense of awe belonging [to] underlying structure and coherence of totality of things' (#46) or 'the experience of transcending one's personal existence on a more cosmic level ... transcending beyond one's body in a communicative way or an emotional way or feeling sort of more, that there is more beyond oneself' (#79).

19. Similarly, 'if we look at the etymology of the term religion, it comes from the Latin *religio*, *religare*, which means to bind ... and the Hellenistic and especially in the Roman periods, of course, religious ceremony and observance was precisely to bring together a highly ethnically, linguistically, culturally disparate empire into common observance.' (#61)

20. In a similar vein, if 'spirituality is something to do with one's experience and one's disposition towards the world', religion 'is a communal phenomenon with its traditions, its myths, and its practices' (#75).

21. In full: 'I would see myself a Durkheimist on this, that separating religion from society is not possible, so that to be social is also to be religious and vice versa ... there isn't a separate religious sphere but it's all somehow combined, that's an important starting point because it frees you up ... from having to try and identify churches, icons, symbols, ritual behaviour and all that sort of stuff which is very difficult and contentious and takes us back to the unique. Probably not a definition but it's more a kind of position' (#52).

22. In full: 'The deepest desires of humankind are to not die and to feel that the world isn't random and cruel and that they have a kind of father figure who will protect them A, from death, and B, from the random ravages of the world if they appease that figure and that mass desire is projected into the skies by various different cultures as the figure God, or Gods or whatever it might be and that primarily is what religion is' (#86).

23. See discussion in chapter 3 but also, more bluntly, Max Planck's famous statement: 'Anybody who has been seriously engaged in scientific work of any kind realizes that over the entrance to the gates of the temple of science are written the words, Ye must have faith. It is a quality which the scientists cannot dispense with'. Planck, Max, *Where Is Science Going? The Universe in the Light of Modern Physics* (London, 1933), p. 214.

24. In a similar vein, 'I don't think propositions are particularly important. I think it's our relationship with other beings that is religion ... you just have to experience some force that impinges upon you' (#67) and 'through studying religion and non-religion, I actually think it's far less about the cognitive aspects, the propositional beliefs than many people think it is' (#74).

25. In a similar vein, 'Shintoism has no dogmas, the Catholic Church has very many dogmas' (#33) and 'I guess, in Protestantism, there has maybe been a particular focus on propositional beliefs ... but I think if you study a Protestant church in actual fact, it's just as much or more about other dimensions' (#74).

26. In a similar vein, religion meant 'things that cannot be tested and proven, yet people believe them' (#29), '[religion] doesn't actually explain anything or any way in which things happen, which is the complete antithesis of science' (#18), and '[religions is] less about observable proofs, and more about taking a leap of faith about things you cannot see or know with just your senses' (#38).

27. 'I personally tend to think of it as early attempts to explain the world, an incomprehensible world ... Religion properly is the study of the universe and our understanding of the universe.' (#19, #22)

28. For example, 'what I would say about religion is that it has nothing whatever to do with science. It is as different from science as art is' (#26).

6

Disaggregating Science and Religion

Public Views

Introduction

In his book *Morals Not Knowledge: Recasting the Contemporary U.S. Conflict between Religion and Science*, John Evans attempts to dislodge the idea that there is a 'foundational conflict' between the two in the public mind. Central to his case is the recognition that the discourse around science and religion is conducted by human beings, rather than in the abstract, and that human beings are always located in and shaped by specific contexts, concerns, and objectives. In the light of this, Evans' argument separates what he calls 'elite' discourse on science and religion from 'public' or 'popular' discourse. If we are serious about understanding the relationship between science and religion, it is important to know what 'the public' is talking about when they talk about each, and not just those whose life and work they are.

Evans defines the elite as those who have a social role that allows them to influence the views of those beyond immediate acquaintances and family members, and he singles out the elite operating within academic circles, who have (or at least pride themselves in believing that they have) a reasonable, justified, structured, and systematic worldview. In his view, this elite engages with questions of science and religion within a framework of warranted truth claims and, as a result, they tend to focus the science and religion debate on epistemological considerations.

By contrast, he contends that the wider public, while having (sometimes very strong) views on science and religion, 'do not have the time, motivation, or desire to make their beliefs logically coherent' in the way that the elite approach demands.[1] It is not that knowledge or coherence is irrelevant to them; rather it is that they view the science–religion interaction primarily through the lens of how they, and how wider society, live and work.

For this reason, Evans argues, the *popular* focus tends to be on moral rather than epistemological considerations, such as questions of (1) which institutions ('religious' or 'scientific') get to set the moral purpose and meaning of a society (2) what implicit moral ideas are embedded in scientific claims, and particularly those scientific claims, like Darwinism, that have implications for the nature of the human; and (3) how medical technology, such as that dealing with embryonic stem cells or genetic modification, should be used and regulated.

There is certainly something to be said for this division, although, as we shall see, it is important not to draw the line too definitively. Evans himself recognises that elites do engage with the moral dimension of the science–religion debate, and that 'popular' views do encompass epistemological considerations, albeit not in those terms. The point is not so much that elite and public views on this issue operate in hermetically sealed conceptual categories, so much as that there are varying views on the issue, depending on the social (and educational) 'location' of the people involved.

In spite of what one of our expert interviewees remarked—'I have to say gathering the views of lots of uninformed people obviously isn't going to lead to much enlightenment' (#33)—*public* opinion, on science, on religion, and on the relationship between the two, does matter, particularly if we are to take seriously the idea of meaning residing in usage. However unsystematised, inarticulate, confused, or sometimes downright wrong it may sometimes be, popular opinion on science and religion cannot simply be ignored.

The contours of public understanding of science and of religion

Before attempting to get a handle on what 'the general public' thinks science and religion actually are, it is worth tracing the contours of that public and, in particular, how religious and how scientific it is, where it draws its views from, and how confident and knowledgeable it is. By doing this we will question the idea that there is such a thing as *the* general public on this matter, and draw out the different ways in which popular opinion conceives science and religion.

Religious demographics

According to the 2021 Census, 37% of the population of England and Wales have no religion, 46% are Christian, 6.5% Muslim, 1.7% Hindu, and less

114 THE LANDSCAPES OF SCIENCE AND RELIGION

than 1% each Buddhist, Jewish, and Sikh.[2] There is an evergreen debate over the meaning of these data, with many people, on the basis of opinion polling, arguing that the Census inflates the number of people who claim the 'Christian' label at the expense of 'no religion'. According to the Theos/Faraday/YouGov survey, which was in the field at around the same time as the Census, 53% of UK adults said they were non-religious compared to 35% Christian.[3]

Whichever dataset is more accurate, it is clear that there is considerable divergence between claiming a religious, particularly Christian, identity and engaging in religious practices. For example, the Theos/Faraday/YouGov data showed that 33% of people who say they are Christians never pray, 56% never or practically never attend a place of worship, and 69% never or practically never read holy texts.

In effect, the UK population is majority non-religious in terms of belief, practice, and, increasingly, affiliation, a fact that has some relevance when we come to consider how popular opinion on religion is formed. Unlike a few generations ago, where a majority of people either practised a religious faith themselves or were brought up in a religiously practising household, today a declining minority of people will have first-hand experience of religious faith.[4] Their understanding of what religion is will be almost exclusively second-hand.

Education

Educationally, the data also point to personal unfamiliarity with both science and religion. According to the 2021 Census, almost exactly a third (33.8%) of adults in England and Wales achieved a terminal Level 4 in education, meaning Higher National Certificate, Higher National Diploma, bachelor's degree, or post-graduate qualification.[5] Given that about 45% of degrees are currently in a science, technology, engineering, and mathematics (STEM) subject,[6] this means that around 15% of the adult population, or 7.4 million people, have some kind of 'high-level' science-based qualification. The Theos/Faraday/YouGov study was roughly in line with this, reporting that 14% of people had an undergraduate degree or technical qualification in science, 5% a master's degree, and 2% a doctorate.

By contrast, the humanities have experienced a long-term relative decline in UK universities, with the total number of humanities students at UK universities falling by around 40,000 in the 2010s. By the end of that decade, the humanities accounted for around 8% of degrees, compared with 28% 50 years earlier.[7] Religious studies, of any description, forms only a minority of that category, and according to the Theos/Faraday/YouGov study, 2% of the population has an undergraduate degree, 1% a master's degree, and less than 1% a doctorate in a religion-related subject.

What this means is that the terminal level of education for the majority of the UK population is either A-level/Scottish Highers/International Baccalaureate (IB) (16% for science; 5% for religious studies), GCSE/Scottish Standard Grade (40% for science; 32% for religious studies), or no qualification at all (20% for science; 55% for religious studies).[8] In other words, and in a similar vein to the demographic data, for the vast majority of the UK public, their main source of information about science or religion will not come from first-hand educational experience. Unburdened by personal, background, household, or educational experience after the age of 18, the majority of the public know what they know about science and religion via the media.

Media

According to the 2014 Public Attitudes to Science (PAS) survey, conducted by Ipsos MORI, 42% of the UK population said that they learned about new scientific research findings from TV news programmes, 26% from other TV programmes, 23% from print newspapers, 16% from on-line news, and 9% from radio news.[9] When repeated five years later, the PAS 2019 survey registered a shift from print to on-line, with 13% finding out about science from print newspapers, compared to 28% from on-line news websites, and 15% from Facebook. However much these figures change over the coming years—and there can be little doubt that the trend towards on-line sources will continue, with all the attendant problems about misinformation—the key point is that for the majority of people, science is something they encounter primarily embedded in a (newsworthy) narrative.

In line with Evans' work in the US, public conceptions of science in the UK are tied up with science *doing* things—warning, healing, predicting— many of which have a wider social, political, moral, environmental, or personal significance. To take just a few examples from the week this chapter was being drafted: 'Scientists warn of crop failure "uncertainties" as Earth heats up'[10]; '"A huge relief": scientists react to hopes of UK rejoining EU Horizon scheme'[11]; 'How playing board games can make your child perform better at school, according to scientists'[12]; 'Science has discovered when we hit our physical and mental peaks ... and the results show youth isn't always an advantage'[13].

To be clear, the moral or personal salience of a science news story is not always uncomplicatedly positive. Media headlines also report 'Lab-grown meat: the science of turning cells into steaks and nuggets'[14] or 'Why scientists are killing millions of Chinese mitten crabs'.[15] Moreover, not all science-in-the-news narratives have an obvious social or personal salience, with many being straightforwardly about new discoveries, for example 'Scientists witness early universe in slow-motion for first time',[16] 'Scientists use deepfake AI images to understand the sun's atmosphere',[17] or 'Scientists say they've finally worked out what's causing [an enormous gravity hole in the middle of the Indian Ocean]'.[18] Even here, however, where the emphasis is apparently more about knowledge than about morals, science is embedded in a narrative of discovery, understanding, and insight. It is still part of an epistemologically progressive, and usually if not always a morally or existentially progressive, narrative.

Data for religion and media are harder to come by. The 2022 HarrisX Global Faith and Media Study reported 'universal agreement' among journalists that coverage of religion had become 'more marginalized due to a set of newsroom dynamics', including squeezed budgets, lack of specialists, lack of representation among staff, and anxiety around the subject in an era of religious politicisation and polarisation, alongside a sense that religion is not seen as a driver for reader engagement 'unless [it] correspond[s] to a narrative of controversy', and a sense that religion is positioned 'as a conservative or extreme force in coverage'.[19] The result is a pervasive sense of frustration, among journalists, regarding the coverage of religion.

That same sense of frustration was evident among the general public, 81% of whom in the study claimed that religion media coverage perpetuated the stereotype that most religions either were against homosexuality or abused children, or promoted radicalism, or hindered women. Furthermore, 43% claimed that the media's current approach to religion created unease and anxiety, especially among secular audiences.

The HarrisX study covered 18 countries, among them the UK, but it is not clear how far these points applied specifically to the UK audience. Moreover, there are several important caveats to note in all this. First, the claim that the media misrepresents an interest group is hardly an unfamiliar one; rarely if ever do we hear praise for media coverage of this nature. Second, it is transparently not the case that *generic* coverage of 'religion' is uniformly hostile, dismissive, or even controversial. A similar survey of news headlines shows that the term 'religion' is not especially common a term and not used negatively when it is.[20]

That recognised, the cultural dominance of stories connecting *particular* religions with *particular* problems—Islam with radicalism, violence, ill-treatment of women, and terrorism; Catholicism with child sex abuse and institutional cover-ups; Anglicanism with conservative views on homosexuality, and with division within the Communion; evangelicalism with rejection of evolution and the support of Donald Trump; Hinduism with aggressive nationalism in India; Buddhism with the mistreatment of minorities in Myanmar, etc.—not only ensures that religions, like science, are encountered as embedded within a newsworthy narrative, but also that, unlike science, this narrative is frequently a problematic or even threatening one. To a public that is largely devoid of first-hand experience of religions, and therefore heavily reliant on mediated information for the formation of their views, this is incalculably important.

Knowledge

Encountering science and religion primarily through a (morally and socially) freighted media narrative does not necessarily equate to ignorance. When asked, in PAS 2019, how well informed they felt 'about science, and scientific research and developments', half of UK adults said either 'very well informed' (7%) or 'fairly well informed' (43%), while half said either 'not very well informed' (41%) or 'not at all informed' (9%).[21] There is apparently little sense of insecurity or ignorance amongst the UK public concerning its level of science knowledge. Only a quarter of people (26%) said they did not feel clever enough to understand science, and the proportion who thought that they heard or saw *enough* information about science (44%) matched the proportion who said they heard or saw too little (47%).

What this relatively confident (or at least balanced) self-perception means for the actual level of public science knowledge is a bit more complex. Both the Theos/Faraday/YouGov study and PAS 2019 attempted to get a basic understanding of people's level of science knowledge by presenting them with a series of factual statements, such as 'All plants and animals have DNA' or 'Lasers work by focusing sound waves', with which they could either agree or disagree, thereby giving a general idea of how knowledgeable they were.[22]

118   THE LANDSCAPES OF SCIENCE AND RELIGION

The results were mixed. PAS 2019 found that 93% of people correctly believed that 'all plants and animals have DNA',[23] 70% of people correctly disagreed that 'all radioactivity is man made',[24] 55% correctly knew that 'electrons are smaller than atoms',[25] and 37% of people correctly believed that 'one kilogram of lead has the same mass on earth as it does on the moon'.[26] At a summary level, both the Theos/Faraday/YouGov study and PAS 2019 split the sample into three groups—high, medium, and low knowledge—based on their overall scoring.[27] According to this classification, PAS found that a quarter (27%) of their sample reported low levels of science knowledge, just over half (53%) were classed as medium, and a fifth (20%) had high knowledge. Whether you think these results are good or bad will largely depend on your expectations.

The Theos/Faraday/YouGov research went a bit further here. Its division of science knowledge was slightly different in that it simply split the sample into terciles. The study then compared each tercile with a corresponding tercile pertaining to levels of science *confidence*—how confident did people feel in their level of scientific knowledge?—thereby allowing it to assess the extent to which confidence about science mapped on to actual knowledge about science.[28] The results (shown in Table 6.1) demonstrated a broad, but hardly overwhelming, correspondence.

Around half those who had low confidence in their level of science knowledge did in fact demonstrate a comparably low level of science knowledge. Over half of those who claimed high confidence in their science knowledge demonstrated a correspondingly high level of knowledge. However, 15% of those who were lacking in confidence showed high levels of knowledge (luck? modesty?); about the same proportion who had high confidence

Table 6.1 Cross tabulation of science confidence and science knowledge.

		Science confidence		
		Low	Medium	High
Science	Low	49%	27%	16%
knowledge	Medium	36%	34%	26%
	High	15%	39%	58%

Source: Theos/Faraday/YouGov 2021; Q6 v. Q14.

came in the bottom tercile for knowledge (arrogance? bluffing?); and there was effectively no correlation between those who displayed medium levels of science confidence and their science knowledge.

The Theos/Faraday/YouGov survey conducted a similar exercise when it came to religion, presenting respondents with a quiz to determine levels of knowledge, although this time without a comparison question about confidence.[29] The results were weaker than those for science. Some questions were well answered; 49% of people correctly agreed that 'Muslims believe that the Qur'an was revealed to Mohammed by an angel', 68% correctly agreed that 'Buddha means "enlightened one"', and 70% correctly agreed that 'Hindus believe in reincarnation'. However, 42% incorrectly thought that 'Christians believe that Jesus was resurrected on Good Friday' and 84% incorrectly thought that 'The Immaculate Conception refers to the belief that Mary was a virgin when she gave birth to Jesus'. Overall, however, uncertainty seemed to be the default option when it came to religion, with 41% of people saying they didn't know whether 'the holiest part of the Jewish Bible is called the Torah', 51% not knowing whether 'Jews worship Moses as divine', and 55% not knowing whether 'King Solomon was the son of King David in the Bible'.

The two sets of questions are, of course, ultimately incommensurable, and it is impossible to say whether one was more challenging that the other. There is no way of comparing whether it is harder to know whether 'the Qur'an is written in Arabic' than whether 'lasers work by focusing sound waves'. However, the data do show that the average number of 'don't knows' for the science quiz was 21% per question whereas it was 33% for the religion one, and they also show that the mean religion score was 3.6 (out of a possible 18) compared to 8.5 for science.[30] In light of this, it seems reasonable to conclude that basic public knowledge of religion is weaker than science. Without much personal or educational knowledge, and ill-served by the mediated stories of religion, the UK public is less confident and more erroneous in its basic understanding of religion.

Conclusion

A number of things emerge from this discussion of the contours of the 'public': first, public understanding is largely second-hand, formed less through personal or educational experience than it is through media consumption; second, media consumption is increasingly shifting towards a social media

120 THE LANDSCAPES OF SCIENCE AND RELIGION

basis, for which there is (even) less research, but the reliability and veracity of which rarely leaves people impressed; and third, mediated consumption embeds the encounter with science and with religion in narratives that are commonly weighted with significant social, moral, personal, or epistemological significance. Science and religion are not experienced as things-in-themselves (whatever that might mean) but as things that heal or harm, develop or destroy, save or end human life and communities. Finally, the narratives in which science is commonly embedded are usually (though not exclusively) epistemologically, technologically, morally, and personally progressive, centring on science's ever-growing ability to know, control, decide, and improve on the material world, whereas those in which religions are embedded are usually (though not exclusively) socially and ethically problematic.

Public understanding of (what) science (is)

In both 2014 and 2019, the PAS survey asked respondents what came to mind 'when I talk about "science"'.[31] The wide range of words and phrases offered by respondents precludes neat classification, but certain patterns and groupings did emerge. In the first instance, and in both waves, the most common association with science was, in effect, being taught it at school. Some combination of biology/chemistry/physics or 'school' itself (or for a minority 'disliked at school' or 'horrible teacher'!) was the association mentioned by 41% of people.[32]

Among those for whom the idea of science went beyond their education, the four most common sets of associations each placed science within a particular social and/or progressive narrative. First, there was the connection with medicine and healthcare. In PAS 2019, 16% of people connected science with 'health/drugs/cures for diseases/hospitals/doctors/medicine/hygiene'. Second, there was the idea of exploring space. In PAS 2019, 15% connected science with 'space/rockets/astronomy', a category that encompasses both Brian Cox-style programmes about the universe and Elon Musk-style ambitions to send rockets to and colonise other planets. Third, there was technology. In PAS 2019, 13% of respondents associated science with 'technology' and a further 2% with 'new appliances/new technology'.[33] Finally, there was a connection with what one might call a straightforwardly progressive understanding. In PAS 2019, 10% associated science with some combination of 'advancement/progress/the future/better world/helping mankind/easier living'. According to these four understandings, science—at least when it is considered beyond school level—is something that warns, heals, explores, develops, improves, and perhaps even saves humanity.

However strong this understanding is, it would be a mistake to imagine that it is an alternative to a more epistemological or methodological understanding of science. According to PAS 2019, 15% of people instinctively connected science with 'ideas/innovation/invention/discovery/research/analysis/logic' (11% did so in PAS 2014), 7% associated it with 'experiment/inquisitive/understanding' (10% in PAS 2104), and a further 4% connected it with 'learn/study/knowledge'.[34]

This is an important if easily overlooked point. The epistemological conceptualisation of science is not top-of-mind for the general public, among whom it takes a secondary role behind a more social and progressive understanding. However, that secondary role is, crucially, facilitatory. It is precisely by being so competent, epistemologically speaking, that science is so securely located within a progressive agenda. Put another way, it is only by knowing so well, that it can warn, heal, explore, develop, improve, and save.

This was shown in the Theos/Faraday/YouGov survey which asked people directly about their epistemological perceptions of science and showed that the majority thought that science was a uniquely reliable method for acquiring knowledge. When asked whether they thought that 'science is the only way of getting reliable knowledge about the world', 54% of UK adults either agreed or strongly agreed, compared with 20% who disagreed or disagreed strongly. In a similar way, when asked whether they thought 'science is less affected by bias and error than other human activities', 45% of UK adults either agreed or strongly agreed, compared with 19% who disagreed or disagreed strongly. Approaching the same issue from the opposite position, when asked whether they thought that 'science ultimately needs faith to work', far more people (50%) disagreed or strongly disagreed than agreed or strongly agreed (17%). In other words, science's uniquely reliable epistemic status lies close to the root of people's belief in the progressive function it performs in society.

This strong appreciation of science's ability to acquire reliable knowledge may surprise those who have read a lot of stories about anti-vaxxers, or monitored US public opinion on science, or imbibed the idea that there is a crisis of trust in science among the US (and indeed UK) public. Is not the same scepticism that is directed towards scientists in America[35], and

122 THE LANDSCAPES OF SCIENCE AND RELIGION

indeed towards almost every public institution in the UK, not also directed to *science* and *scientists* in the UK?

The answer appears to be no. Trust in science and scientists remains high. Indeed, Covid-19 appears to have improved it. A study conducted in early 2023 reported that more than a third of people said that their trust in science, in particular genetics, increased during the pandemic.[36] The Theos/Faraday/YouGov quantitative research was conducted during the pandemic (May–June 2021—saliently, after a vaccine had been developed and during the time it was being successfully implemented in the UK) and perhaps predictably reported that 65% of people disagree or strongly disagree that 'the dangers of science outweigh its benefits', compared with 9% who agree or strongly agree. Similarly, when asked whether they thought 'science changes too much to be completely trustworthy', 53% of UK adults disagreed or strongly disagreed, compared to 16% who agreed or strongly agreed. Science is trusted to do what science claims to be able to do.

That recognised, there are some important caveats to this. First, some sectors of the public are somewhat more sceptical than others, an issue to which we return in the Conclusion.

Second, the political/social/economic location of the science and the scientists in question makes a difference. According to PAS 2019, around 90% of the public trust scientists, engineers, and researchers working for universities. That falls to around 80% for those scientists working for charities or environmental groups, 75% for those working for government, and 57% for scientists who are working in the private sector. In the same way as science is encountered by many people primarily embedded in socially salient narratives, its status and credibility is tied up with the particular social, political, and economic institutions in which it operates.

Third, the *kind* of scientific activity in question elicits different levels of trust and there are some areas of scientific research and progress about which the general public is distinctly more sceptical. For example, 41% of people on balance think the benefits of nanotechnology outweigh the risks (7% think the reverse), 70% of people think the benefits of stem cell research outweigh the risks (7% think the reverse), and 86% think the benefits of vaccination research outweigh the risks (4% think the reverse). By contrast, the public is much more hesitant about animal testing for medical research,[37] genetically modified crops,[38] and nuclear power.[39]

Just because science is uniquely well qualified, epistemologically and methodologically speaking, to acquire reliable knowledge, and therefore closely associated with the narrative of human progress, that doesn't mean that all scientific locations or activities are equally appreciated or trusted. Nor does it mean that science in general is therefore unchallengeable or omnicompetent. The nature of the public's reservation with, and sense of the limitations of, science is relevant here, if not always clear or consistent.

When people were asked whether they thought that 'science can answer big questions about life and meaning', 54% agreed or strongly agreed that it could, compared to only 14% who disagreed or disagreed strongly. Quantitative research does not afford the opportunity to interrogate how respondents interpreted 'big questions about life and meaning', but on the surface at least this looks as if a majority of people see science's purview and competence being metaphysical as well as just physical.

That said, when they were asked whether they thought 'there are some things science will never be able to explain', 64% of people agreed or strongly agreed, compared to 15% who disagreed or strongly disagreed. In a similar vein, 57% of people agreed or strongly agreed that 'science is only able to explain part of reality', compared with 18% who disagreed or strongly disagreed. And again, while 29% of people agreed or strongly agreed that 'science will be able to explain everything one day', slightly more (40%) disagreed or strongly disagreed with the statement. However reliable science was understood to be as a means of acquiring knowledge about the physical world, and perhaps even about the big (metaphysical) questions, it was nonetheless apparently limited in its scope and ultimate potential.

Where science's perceived limitation *could* be pinned down was in its relation to ethics. It is important to be clear about this. People do not think science or scientists are unethical. PAS 2019 reported that only 9% of people thought scientists dishonest. The survey showed that the public placed emphasis on scientists' need to be 'ethical', 'honest', and 'open-minded', and reported that 66% of people felt that scientists should spend *more* time discussing the social and ethical implications of their work with the public. In other words, it is certainly not the case that ethics has nothing to contribute to science, or vice versa. When the Theos/Faraday/YouGov survey put the statement 'science has nothing to say about ethics' to the public, rather more people disagreed or disagreed strongly (38%) than agreed or agreed strongly (24%).

Nevertheless, having an ethical salience and an ethical role is not the same as being ethically sufficient. Put another way, science's impressive record in knowing about reality and in powering human progress does not mean it is competent to steer that progress, let alone to adjudicate on personal ethics. When asked whether they agreed that 'science cannot tell you

124 THE LANDSCAPES OF SCIENCE AND RELIGION

how to live your life', considerably more people agreed or agreed strongly (55%) than disagreed or strongly disagreed (17%). A majority of people (56%), according to PAS 2019, believe that scientific activity needs to be regulated by government or parliament, almost twice as many as those who think it should be regulated by scientists themselves (31%). And when asked whether they thought 'science can sometimes damage people's sense of morality', more people (36%) agreed or agreed strongly than disagreed or disagreed strongly (30%).

Such data cannot be understood outside of the—at the time very live—debate about the extent to which governments should simply be following scientific advice when it came to dealing with the Covid-19 pandemic. At the time of polling, scientific communities across the world had covered themselves in glory by conceiving, preparing, testing, and producing a vaccine with astonishing speed and, it seemed, success. However, the UK was also just emerging from the third national lockdown, debates around the necessity and efficacy of which were growing. The phrase 'follow the science' was popularly deployed but widely contested. On the one hand, Anthony Fauci, chief scientific spokesperson for the Biden administration and possibly the world's most prominent scientific figure at the time, commented in an interview published in *The Atlantic* that 'there's a full commitment on the part of the administration to let scientific principles be the sole guide of what we do. Absolutely, the underlying core basis of what we do is all science'. On the other hand, as the *Journal of Medical Ethics* put it, 'with all due respect to Dr. Fauci ... while scientific principles and evidence should guide policy decisions in response to the twists and turns of the pandemic, they can never be the sole basis for these decisions'.[40]

In this way and at the time of our survey, science's epistemological and methodological supremacy in the public's mind, combined with its close association with progressive narratives, particularly those concerning health and medicine, was coming into very public contact—and occasional tension—with science's apparent ethical limitations.

Public understanding of (what) religion (is)

In imitation of PAS 2014 and 2019, Theos/Faraday/YouGov commissioned YouGov to put to the public the same generic question about religion as PAS did about science: 'When you think about religion what comes to your mind?'[41] The question was open-ended and multi-coded so respondents were limited neither to pre-chosen categories nor to a single association. The results map broadly onto a number of the 'family features' of religion identified in our elite interviews and our examination of various legal, anthropological, and philosophical 'definitions' of religion.

Firstly, the most common single response to the question was 'belief', given by 23% of respondents. In addition to this, a little way down the list was 'faith', named by 11%. This clearly reflects the fact that in contemporary English, the words 'religion', 'belief', and 'faith' are often used interchangeably, almost as synonyms, in public discourse. A religion is a 'faith group', a religious individual is a 'person of faith', and they are 'a believer', which seems to imply, contentiously, that non-religious people have neither faith nor beliefs.

That being so, it might justifiably be said that the word association here is just that—an involuntary association with another culturally coterminous term. Even if that is the case, however, it is still salient. The fact that belief/faith so readily came to mind is indicative of the way in which, in our current culture, the methodological/epistemological features of 'religion' have (for whatever reason) risen to the surface. Whether they are conscious of it or not, many people conceive religion primarily in these terms. Religion is associated not only with *what* people believe but also, perhaps primarily, with *how* they acquire and hold those beliefs. This is a critical point, to which we shall return in our discussion of methodology in part II

Secondly comes a caveat to this point. The third most common association (we shall return to the second most common one presently) was with 'God', named by 15% of people. A further 1% said 'higher power' or 'supreme being', 4% said Jesus, and specific religions ('Christianity/Judaism/Islam/Hinduism/Sikhism') were cited by a further 12% of the sample. In other words, although religion is understood in methodological/epistemological terms, it is also conceived as having content. It *is* about 'God', however that term may be interpreted, and presumably also about the teachings inherent in specific religions. Religions can be defined as being about believing, but they can also be defined by what they believe in. Content matters.

Thirdly, that content may not be what religious believers think it is— certainly not what they would like it to be. The second most common word that came to mind when people were asked to think about religion was 'war', cited by 19% of people, only slightly fewer than mentioned 'belief'. Curiously, this was much less of an issue for younger respondents, with 8% of 18–24 year olds and 11% of 25–34 year olds making this connection, compared

126   THE LANDSCAPES OF SCIENCE AND RELIGION

with 23% of over 45s. No such similarly sharp age difference was visible with other associations, suggesting that this link may possibly be fading. Nevertheless, it remains strong today, with a further 2% of people saying 'hatred' and 1% 'cause of problems'. The straightforward association of religion with war is a very old trope, but one that is clearly alive and kicking, one of the bluntest indicators of regressive narrative in which the public understanding of religion is embedded.

Two other associations with religion are worth highlighting. Religion is a social/cultural phenomenon: 5% of people connected religion with 'support/community' and another 5% with 'culture/tradition/civilisation'. In addition to this, 13% of people said 'church/temple/mosque', which, however prosaic an association it is, still draws attention to the physical manifestation and presence of that religious practice in the wider society.

Finally, there was a clear moral conception of religion, although one that, as with the instinctive association of religion with violence, had a very dark shadow to it. Only small numbers of people, 1% in each case, associated religion with 'values' or with 'morals'. By comparison, more people associated it with restrictive forms of morality (3% with 'rules/guidelines/restrictions'), or outright oppressive forms (3% 'control/subservience'), or antiquated ones (3% 'old fashioned/boring/outdated'), or simply hypocritical ones (8% 'hypocrisy'). Given that the more positive morally loaded terms ('love/peace/heaven') amounted to a mere 4%, this is a rather depressing outlook for those sympathetic to religion, and one that further underlines the deep and heavily mediated association of religion with a regressive and dehumanising presence in the world.

This exercise in unprompted association draws out the way in which certain features of the family definition of religion—in particular, epistemological, social, and (un)ethical—are present within broad public opinion. This picture is filled out further by the prompted questions on this subject we put to respondents. We presented people with four statements about religion that encompassed four different ways of understanding what it is. These were the metaphysical/supernatural ('religion is a set of beliefs about spiritual issues (e.g. God, the soul))'; the cultural/ritual ('religion is a set of cultural practices and rituals'); the ethical ('religion is a set of beliefs about morality'); and the (proto-) scientific ('religion is a set of beliefs about the physical world (e.g. nature and the universe))'.[42]

Respondents were able to tick as many as they agreed with but were also asked, if they agreed with more than one, which one they agreed with most strongly. In the first instance, opinions were quite widely spread. Three-quarters (76%) of people thought that religion was about spiritual issues, three-fifths (60%) thought it was a set of cultural practices and rituals, half (50%) that it was a set of beliefs about morality, and a third (32%) that it was a set of beliefs about the physical world.

This suggests that, as a general rule, the public considers religion to be a broad entity, encompassing a range of different identities and functions, though more likely to be metaphysical and cultural, than ethical or proto-scientific. However, the picture changes slightly when narrowed to examine which category *best* captured respondents' understanding of religion. The idea that religion is a set of beliefs about spiritual issues remained predominant, with 46% of people thinking that is the best single conceptualisation of what religion is. The figure was even higher for those who considered themselves to be religious. This result appears to contradict any idea that the public thinks religion is only a ritual or cultural phenomenon.

By contrast, whereas, as we have seen, large numbers of people think religion is a set of cultural practices and rituals or a set of beliefs about morality, far fewer people thought these are the *best* understanding of religion. Nine per cent of people thought the ritual/cultural understanding of religion was the most accurate, and only 3% thought that of the ethical understanding. In the public mind, at least, while both of these are clearly features of religion, neither is dominant, both being essentially secondary to, or parasitic on, a different understanding.

More surprisingly, the same proportion of people who think that religion is 'a set of beliefs about the physical world' (32%) think this is the most accurate view of what religion is.[43] In other words, those who think religion exhibits the feature of being, in essence, a protoscientific entity, a set of beliefs about the physical world, strongly tend to see this as *the* feature of religion rather than just a feature. If you think religion is protoscience, you think it is primarily protoscience.

This assessment of the public understanding of (what) religion (is) suggests that that understanding is complex and varied, encompassing metaphysical and epistemological features (albeit not in those words) alongside cultural and ethics ones. Evans is right to draw out the way in which these latter dimensions feature heavily in the popular (US) conceptualisations of religion, but it is important to recognise that their prevalence does not come out of nowhere. In the popular mind, as in the elite one, religion is a criss-cross of different features, and while that definitely includes moral and ritual features—the science-as-progress narrative is deeply embedded

128 THE LANDSCAPES OF SCIENCE AND RELIGION

in the popular consciousness—it would be wrong to imagine that the more 'elite' features of methodology and metaphysics are absent. They may not be expressed in the same terms as professional scientists, philosophers, and theologians use, but they are still present and relevant.

Conclusion

If the nature of the science and religion conflict, such as it is, is predicated on what we understand science and religion to be, it is important to grapple with the thorny question of what 'the public' thinks about this issue. That may not 'lead to much enlightenment', as our expert interviewee said, but it will help us get a handle on what exactly (the majority of) people are disagreeing about when they disagree about science and religion.

The answer shows as much complexity as that which we glean from the professionals. Popular opinion may not fixate on methodological (or epistemological) issues, as Evans rightly argues, but that does not mean it is indifferent to them. Science is a (uniquely) reliable way of ascertaining facts about the world as far as the general public is concerned. However little they may know or care about the scientific method (or methods), it is this conviction that lies at the root of their understanding of science.

Importantly, this understanding is not the end of the matter for them. The general public tends rarely to encounter science 'in the raw'—through working in laboratories, attending conferences, or reading science journals, for example—but rather finds it embedded in narrative forms, many of which are news-based, have a personal or social salience, and are freighted with ethical and progressive considerations. In this way, the public understanding of science often has the same rich, complex, multi-dimensional character we discerned in the elite understanding, even if (once again) that popular understanding has neither the framework nor the vocabulary for it.

The story for religion is both similar and different. One similarity lies in the fact that for many people, religion is synonymous primarily with 'belief' or 'faith'; it is, in effect, a methodological (or epistemological) understanding which places religion firmly at odds with science's more impressive methodological credentials. For all that this tension emerged as a foundational one among elite interviewees, it was nonetheless present in an inchoate way with the general public.

A second similarity lies in the fact that religion, like science, was encountered in a morally and socially weighted narrative. The (very considerable)

DISAGGREGATING SCIENCE AND RELIGION   129

difference here was that that narrative was not progressive or emancipatory, as it commonly was for science, but disagreeable and destructive. War, rules, restrictions, control, subservience, old-fashioned, outdated, and hypocritical were all unprompted associations with the term religion. However else these may be understood, it is not positively.

In this way, the public understanding of religion was rich and complex in the same way as it was for science. Religion had an important methodological dimension or feature, as well as a metaphysical one, articulated as the conviction that it was a set of beliefs about 'spiritual issues' like 'God and the soul'. But it also had pronounced social/communal and ritual features, with significant ethical connotations. In short, all the family features of religion identified in the elite interviews and legal and philosophical definition of religion were visible within the general opinion, albeit, as John Evans correctly identified, in a different balance.

Notes

1. Evans, *Morals Not Knowledge*, p. 10.
2. For the 2021 Census in England and Wales see https://www.ons.gov.uk/peoplepopulation andcommunity/culturalidentity/religion/bulletins/religionenglandandwales/ census2021. The Scottish Census was conducted in 2022, so data were not available at the time of writing. The 2011 Scottish Census question on religion reported 37% no religion, 54% Christian, and 1.4% Muslim. The 2021 Northern Ireland Census reported 17% no religion, 42% Catholic, 17% Presbyterian, 12% Church of Ireland, and 9% other Christian denominations.
3. Data are available at www.theosthinktank.co.uk.
4. To be precise, according to the Theos/ Faraday/ YouGov data, while 48% of over 70s were brought up in practising Christian households, 32% of those in their 50s were, and only 19% of those aged 16–29 were.
5. Education, England and Wales: Census 2021 (https://www.ons.gov.uk/peoplepopulation andcommunity/educationandchildcare/bulletins/educationenglandandwales/ census2021).
6. Higher Education Student Statistics: UK, 2020/21—Subjects studied (https://www.hesa. ac.uk/news/25-01-2022/sb262-higher-education-student-statistics/subjects).
7. Improving the fortunes of the humanities means thinking about post-16 qualifications (https://www.hepi.ac.uk/2021/09/23/improving-the-fortunes-of-the-humanities- means-thinking-about-post-16-qualifications/#:~:text=Improving%20the%20fortunes %20of%20the%20humanities%20means%20thinking%20about%20post%2D16% 20qualifications,-23%20September%202021&text=The%20humanities%20are% 20often%20said,%2C%20graduate%20employment%2C%20and%20funding).
8. Data are from our study, which was UK-based, rather than England and Wales. A-level here included data for Scottish Highers and the IB; GCSEs included Scottish Standard Grades.
9. The precise wording of the question was 'Outside of any formal science lessons or classes you may take, from which one or two of these, if any, do you hear or read about new scientific research findings most often?'.

130 THE LANDSCAPES OF SCIENCE AND RELIGION

10. https://www.aljazeera.com/news/2023/7/4/scientists-warn-of-crop-failure-uncertainties-as-earth-heats-up#:~:text=A%20new%20study%20has%20highlighted, dystopian%20future%E2%80%9D%20without%20immediate%20action.
11. https://www.theguardian.com/science/2023/jul/06/scientists-react-to-news-of-uk-rejoining-horizon.
12. https://www.dailymail.co.uk/sciencetech/article-12271551/How-playing-board-games-make-child-perform-better-school-according-scientists.html.
13. https://www.dailymail.co.uk/health/article-12267117/Scientists-know-hit-physical-mental-prime-not-20s-30s.html.
14. https://www.nature.com/articles/d41586-023-02095-6#:~:text=The%20general%20recipe%20for%20cultured,them%20to%20bind%20into%20fibres.
15. https://edition.cnn.com/world/chinese-mitten-crab-invasive-species-europe-solution-scn-c2e-spc-intl/index.html#:~:text=Its%20omnivorous%20feeding%20habits%20significantly,well%20as%20by%20damaging%20gear.
16. https://news.sky.com/story/scientists-witness-early-universe-in-slow-motion-for-first-time-12913961.
17. https://phys.org/news/2023-07-scientists-deepfake-ai-images-sun.html.
18. https://www.businessinsider.com/scientists-crack-mystery-of-huge-gravity-hole-in-the-indian-ocean-2023-6.
19. https://www.media-diversity.org/resources/the-global-faith-and-media-study/.
20. For example, from the week this was being written: Parliament of the World's Religions hopes to harness faith to address world's ills (https://religionnews.com/2023/08/14/parliament-of-the-worlds-religions-hopes-to-harness-faith-to-address-worlds-problems/); Religion more important to faithful in US than those in UK (https://www.churchtimes.co.uk/articles/2023/11-august/news/world/religion-more-important-to-faithful-in-us-than-those-in-uk); Losing Our Religion review: Trump and the crisis of US Christianity (https://www.theguardian.com/books/2023/aug/13/losing-our-religion-review-trump-crisis-christianity).
21. PAS 2019, Q2.
22. The full list of our statements was 'Electrons are smaller than atoms', 'All radioactivity is man made', 'All plants and animals have DNA', 'More than half of human genes are identical to those of mice', 'The cloning of living things produces genetically identical copies', 'Lasers work by focusing sound waves', 'By eating a genetically modified fruit, a person's genes could also become modified', 'The oxygen we breathe comes from plants', and 'It is the mother's genes that determine the sex of the child'. The PAS 2019 survey added the statement, 'One kilogram of lead has the same mass on earth as it does on the moon'. Our survey adopted a similar approach (using 9 of the 10 questions) but also asked respondents how confident they were in their understanding of various scientific disciplines, which allowed us to ascertain how far the public's self-perception of knowledge (or ignorance) was borne out by reality.
23. We found that 61% of people thought this was definitely the case, and a further 29% thought it probably true.
24. We found 53% were definite about this, and 20% probable.
25. We found that 38% were definite, 17% probable.
26. We did not ask this question.
27. For the PAS study, those who answered five or fewer questions correctly were classified as having low scientific knowledge, those who answered six to eight questions correctly were classified as having medium scientific knowledge, and those answering nine or ten questions correctly were classified as having high scientific knowledge.
28. The confidence scores were drawn from answers to the question 'How confident are you that you understand the basics of the following areas of science?', with the respondents then presented with the following scientific areas: the theory of evolution, the idea of the Big Bang, and the basics of medical science, chemistry, climate change, psychology (the scientific study of the human mind), neuroscience (the scientific study of the

DISAGGREGATING SCIENCE AND RELIGION   131

human brain), ecology (the branch of biology that deals with organisms and their physical surroundings), and geology.

29. The religion statements were 'Muslims believe that the Qur'an was revealed to Mohammed by an angel', 'The Qur'an is written in Arabic', 'Buddha means "enlightened one"', 'Hindus believe in reincarnation', 'King Solomon was the son of King David in the Bible', 'Jews worship Moses as divine', 'The holiest part of the Jewish Bible is called the Torah', 'Christians believe that Jesus was resurrected on Good Friday', and 'The Immaculate Conception refers to the belief that Mary was a virgin when she gave birth to Jesus'.

30. Scoring worked by awarding 2 points if someone answered definitely (true or false) and were correct; 1 of they answered probably (true or false) and were correct; 0 if they were not sure; −1 if they answered probably (true or false) and were incorrect; and −2 if they answered probably (true or false) and were incorrect.

31. In actual fact, the format of the question was changed in 2019. In 2014, interviewers coded open-ended responses to a predefined list, whereas in 2019 respondents were asked a fully open question ('When I talk about "science", please tell me in your own words what comes to mind') and verbatim responses were recorded by the interviewer. This means that findings cannot be directly compared between surveys. The question was open to more than one answer per respondent and thus the total of percentages could exceed 100.

32. In PAS 2014, biology/chemistry/physics was 28%; in PAS 2019 it was 23%. School was 12% in both waves. About 1% of respondents in 2019 nominated 'disliked at school/horrible teacher'.

33. The categories were slightly different in PAS 2014, when 9% made the connection with 'new appliances/new technology' and 1% with 'technology' with no mention of 'new'.

34. Not a category in 2014.

35. https://www.pewresearch.org/science/2022/02/15/americans-trust-in-scientists-other-groups-declines.

36. https://genetics.org.uk/public-perception-of-genetics/.

37. Benefits outweigh risks 42% v. risks outweigh benefits 34%.

38. Benefits outweigh risks 38% v. risks outweigh benefits 33%.

39. Benefits outweigh risks 40% v. risks outweigh benefits 36%.

40. See We can't just follow the science (https://blogs.bmj.com/medical-ethics/2021/12/27/we-cant-just-follow-the-science/); Governments cannot just 'follow the science' on COVID-19 (https://www.nature.com/articles/s41562-020-0894-x); Why 'following the science' to minimize Covid risk may not be so easy (https://www.nytimes.com/2022/02/11/briefing/covid-cdc-follow-the-science.html#:~:text=People%20want%20to%20protect%20themselves,is%20a%20unitary%2C%20omniscient%20force); /); 'Follow the science': as the third year of the pandemic begins, a simple slogan becomes a political weapon (https://www.washingtonpost.com/health/2022/02/11/follow-science-year-3-pandemic-begins-simple-slogan-becomes-political-weapon/).

41. This question was put into the field in September 2023, rather than May 2021.

42. It will be noted that the last of these was only rarely identified as a family feature of religion by elite interviewees, but the popular rhetoric on this topic, particularly in the wake of the New Atheist phenomenon, meant that we thought it appropriate to include it as a feature in this list. The results confirm our inclination.

43. The 10% left outstanding once these four options have been accounted for did not offer an opinion on the topic.

Conclusion to Part I

The hope of identifying a model, or a series of models, or even the absence of a model, for the relationship between science and religion is predicated on having a reasonably clear idea of the two entities involved. Too many attempts to do so either have drawn on implicit and unduly narrow definitions of science and religion, or have basically retrospectively defined them to fit the predetermined model. The way in which so much discussion about (and research on) science and religion focuses on evolution, cosmology, miracles, quantum theory, and Genesis is an example of the first of these. The way in which Stephen Jay Gould defined science (as covering the empirical realm) and religion (as covering the realm of ultimate meaning and moral value) in order to legitimise his 'non-overlapping magisteria' is an example of the second. It is not that the science and religion debate does not incorporate all these different topics. It is that it incorporates more and can't be reduced to these alone.

The need to define science and religion draws us into two separate quagmires, revisiting the demarcation problem for science, and entering a longstanding and effectively intractable debate for religion. These are, however, quagmires from which the Wittgensteinian notions of language as usage and family resemblances can help rescue us. Released from the task of needing to define the terms unambiguously or absolutely, this approach encourages us to observe how the words are used, by drawing on various philosophical, legal, and empirical sources, and, in doing so, to disambiguate the categories in such a way as will expose their various common dimensions, characteristics, or 'family features'.

None of these dimensions is strictly speaking a *necessary* feature, although some are clearly more salient than others. None is sufficient. And no combination of features allows us to draw definitive lines round either category in such a way as would classify one set of things as unambiguously science (or religion) and another set as unambiguously not.

Rather, it leaves us with a fuzzy logic—hard-and-fast categories being replaced by the notion of gradated belonging. By this logic, certain things

CONCLUSION TO PART I   133

(those that exhibit many family features) can be confidently said to be science (or religion). We might call this category (a). Other things—category (c)—that exhibit few or no features, can be confidently said *not* to be science (or religion). But many others—category (b)—exhibit some combination of features, are harder to classify, and are therefore best categorised as partially or contestably science (or religion).

Thus, in (a) we might confidently place molecular biology and chemical engineering, on the one hand, and the celebration of the Eucharist, the Hajj, and Kumbh Mela on the other. In (c) we could reasonably put gossip, gardening, painting, and attending a seminar on Aristotelian metaphysics as falling outside the categories of both science and religion. But (b), and the place of activities in it, is always going to be more contestable, with, for example, hermeneutics, economic history, and pure mathematics arguably belonging in the liminally scientific space, and the opening of parliament, placing flowers and teddy bears at the scenes of accidents, and Wordsworth's experience above Tintern Abbey belonging in the liminally religious space. Readers will have their own examples and may well disagree with these.

The price of embarking on the task of definition is disaggregation, family resemblance, fuzzy logic, and ultimately contestable categorisations. It is, however, a price worth paying, as it leaves us with a much richer, more granular, and more accurate set of ideas with which to understand and map the relationship between science and religion.

Part I of this book has engaged in this process for both science and religion. The process obviously prohibits any simple, dictionary—or even encyclopedia—definition of either category by way of conclusion, and in its place generates a list of dimensions or features that between them characterise the families of science and religion. For science, our combination of philosophical, legal, and empirical sources drew out six; for religion, five. Each of these has a number of sub-categories within it (see Table I.1). The result is a bemusing number of points at which the disaggregated features within the categories of science and religion could come into contact, and therefore in theory tension, or perhaps concord.

Disambiguating science and religion in this way could be accused of muddying the landscapes intolerably—a legitimate criticism but ultimately an irrelevant one given the nature of the entities about which we are speaking. Science and religion are fluid, fuzzy, complex, overlapping, negotiable, socially mediated categories. They do not admit neat definitions, in spite of commonly (if usually implicitly) receiving them in popular discussion or opinion polling.

134 THE LANDSCAPES OF SCIENCE AND RELIGION

Table I.1 Disambiguated 'definitions' of science and religion summarised.

Science ...	Religion ...
is a series of method(s), which involve • formation of questions, hypotheses, models, theories • which are testable • and refutable • by means of structured acts of observation • and ideally experimentation • through which it accumulates evidence • that is amenable to measurement and quantification • and (ideally) mathematical, and open to statistical analysis • and from which it derives conclusions that are provisional • and ideally replicable • and that are used to establish or strengthen further theories on which subsequent hypotheses and observations can be built.	is characterised by the conviction that there is a transcendent dimension to reality, discernible in the existence of • God or gods • or non- or supernatural agency • or a dimension, or entity, or reality • of which humans have (limited and imperfect) knowledge • to which they have (limited and imperfect) access • that may be personal (or described in personal terms) • and is responsible for bringing that immanent world into being • for superintending or governing it • and for showing an on-going interest in it, and especially in human beings.
is the study of • natural or physical or material phenomena • that are in principle observable • and to which humans have common access • and which are amenable to natural(istic) explanation.	is characterised by belief • that takes seriously the truthfulness, relevance, and authority of particular texts, events, and doctrines • and that is both cognitive in nature • but also experiential, affective, and existential • and which takes seriously experiences of awe, wonder, mystification, and love • which help formulate answers to foundational human existential questions, concerning meaning and purpose.

CONCLUSION TO PART I   135

Science ...	Religion ...
is predicated on certain metaphysical presuppositions, such as	involves a personal, practical, material, and bodily response and commitment
<ul><li>the existence of laws of (or immutable patterns or regularities within) nature,</li><li>the intelligibility of the universe (or, at least, its amenability to rational explanation by the human mind),</li><li>the reliability of human senses,</li><li>and the existence of the external world.</li></ul>	<ul><li>that is moral, involving both assent and practice,</li><li>ritual, exercising personal and collective spiritual disciplines,</li><li>and communal, involving commitment to the wider community and its good.</li></ul>
is characterised by the objective of	involves the formalisation or codification of those beliefs
<ul><li>making increasingly accurate and effective predictions about its phenomena of study</li><li>by means of which it is able to facilitate control and modification of the physical world</li><li>which enables human progress and development.</li></ul>	<ul><li>into doctrine, theology, and ethical code</li><li>as a way of structuring beliefs about the transcendent</li><li>and about human life and purpose in the light of the transcendent</li><li>and informing consequent commitments that result from those beliefs.</li></ul>
is characterised by a strong commitment to particular values, including	involves the formalisation or institutionalisation of commitment
<ul><li>curiosity</li><li>epistemic humility</li><li>provisionality</li><li>objectivity</li><li>patience</li><li>honesty</li><li>creativity</li><li>and imagination.</li></ul>	<ul><li>in practices, rituals, patterns of behaviour, and human relationships,</li><li>codifying texts, ethics, laws, membership, and offices,</li><li>and accordingly reclassifying time, space, and material culture in a way that is informed by that ultimate transcendent reality.</li></ul>

Continued

136 THE LANDSCAPES OF SCIENCE AND RELIGION

Table I.1 *Continued*

Science ...	Religion ...
is characterised by its social and institutional nature, being • an inherently communal, collaborative, • and cumulative exercise, • with that communality being structured, formalised, professionalised, and institutionalized, • with its personnel, funding, processes, and conclusions being validated through these institutions.	

The kind of disambiguation we have undertaken in part I, which pays full attention to the various ways in which the words are actually used, is necessary if we hope to pinpoint what exactly it is that people are talking about when they are talking about science and religion, and in particular what they are disagreeing about when they disagree.

In theory, the possibilities—for contact and therefore tension—are legion. If, as the table shows, we can disambiguate science into six family features and religion into five, and if each of those features has discernible sub-categories within them, the different points at which a feature of science could come into potential contact and tension with a feature of religion run into the hundreds.

In reality, the potential for meaningful connection and (perceived) tension is rather more limited. For example, it is hard to see why anyone would see potential tension between science's commitment to measurement and quantification and religion's inclination to ritual or communal practice, or between religion's tendency to codify certain texts, laws, and ethics and science's commitment to making increasingly accurate and effective predictions about the material world. The families of science and religion are entangled, but not at every conceivable point.

Moreover, not every point of contact necessarily constitutes a potential point of tension. Indeed, not only is not every point of contact between the disambiguated family features of science and of religion *not* a point of

CONCLUSION TO PART I   137

tension, but also, to the contrary, a point of contact may underline a particular point of *harmony* between a feature of science and religion. Whether it is regarding metaphysics, in which there is an apparent point of concord between science's presupposition of the lawfulness of the universe and (some) religious belief about the source of those laws (a 'lawgiver'); or regarding ethics, where both science and religion share a deep commitment to the certain values (honesty, co-operation, creativity, etc.) that are fundamental to the practice of both; or regarding public affairs, in which both science and religion have allied in humanising reform, such as concerning public health in the later nineteenth century ... there are plenty of examples in which science and religion have been successful allies rather than antagonists.

All that acknowledged, the focus of this book, in line with so much of the energy in the wider conversation, is on the areas where science and religion are claimed to be tension. It is not accident that the debate on this topic between Alvin Plantinga and Daniel Dennett was published as *Science and religion: are they compatible?* and that even Plantinga's extended argument that they *were* largely compatible was entitled 'Where the conflict really lies'. As any newspaper editor knows, conflict sells.

So it is that, while we can see how the list outlined in the table affords plenty of opportunity for complementary or harmonious contact, and plenty more opportunity for mutual indifference or separation, it does also permit us to explore those areas in which different features (within the disambiguated 'definitions') of science and religion come into apparently conflictual contact. In effect, it allows us to pinpoint what it is that people are actually disagreeing about when they disagree about science and religion. It is to these we turn in part II.

PART II

TOURING THE LANDSCAPES OF SCIENCE AND RELIGION:

Understanding Where the Disagreements Lie

Introduction to Part II

Part I of this book surveyed the landscapes of science and religion by defining and disambiguating the terms themselves. Neither category is tidy, homogenous, well defined, or clearly unified. Rather, each is compound, complex, and sprawling, and comprising various features that are more (or less) important to the category as a whole. Adopting and adapting Peter Harrison's metaphor of territories, we compared each to a landscape, crowded with features that give each its particular characteristic. Thus, whether something can legitimately be called science is answered by the nature of its methods, object of study, presuppositions, objectives, values, and institutions, just as whether something can legitimately be called religion is determined by its attitude to the transcendent, to belief, to personal commitment, and to the formalisation of those beliefs and commitments.

Contrary to Stephen Jay Gould, certain features do appear on both of these landscapes, or put another way, the landscapes (in his terms, the 'magisteria') do overlap. It is not as if humans engage in wholly separate endeavours, one oriented towards establishing facts and the other values, the former falling under the rubric of science, the latter religion. That kind of intellectual demarcation is too simple, too tidy, and fails to correspond to reality. Rather, both science and religion are part of the wider human endeavour of understanding and inhabiting the world, and so their landscapes overlap in many places.

In this regard, the more familiar metaphor of cartography may feel more appropriate here: science and religion both provide maps of the same reality, just different kinds of maps, in the way that an atlas might contain physical and political maps of the same territory. We have resisted this metaphor, however, because the idea of 'intellectual cartography' all too easily assumes that maps are of the same territory. This is, as it were, the error the stands at the opposite extreme to Gould's. Rather than 'non-overlapping', this approach implies that science and religion are wholly coterminous, surveying the same ground just in different ways.

And yet, as we have seen in part I, science and religion are not coterminous. They are not doing, or even attempting to do, the same thing. While they do share many features, some of which we will explore in part II, they are not simply different maps of the same territory. They are different landscapes of human activity.

Threading its way between these two errors, part II of the book turns to some of the 'features' common to each of the 'family definitions' of science and religion (or, to adopt another use of the word, to the 'features' of the different landscapes of science and of religion). In particular, given the overall focus of the book, it explores those features that appear in both landscapes, about which there appears to be genuine and profound disagreement.

As we suggested in the conclusion of part I, the sheer number of features with each of the disambiguated 'family definitions' of science and religion means that there is a potentially vast number of ways in which we might compare the landscapes of science and religion. However, on the basis of our research, four main features emerged as genuine points of potential conflict.

Both science and religion have certain metaphysical commitments that were judged—by some, potentially—to be in tension. Both have approaches to the acquisition, modification, and defence of beliefs and knowledge, what we have called here methodology, that are understood—by some, potentially—to conflict. Both science and religion have or develop understandings of the human nature and condition that are considered—by some, potentially—to be incompatible. And both science and religion bring their perspectives to bear on a whole range of public, social, and ethical issues in a way that frequently, if not necessarily, generates at least the popular impression of conflict.

To return to the arguments of part I, none of these perceived tensions is strictly speaking necessary; there will be plenty of conceptions of religion and of science in which these metaphysical, methodological, anthropological, and social/political features are either not present or not felt to be in tension. But the research for this book repeatedly showed that not only are these legitimate features of both science and religion, but also they are features over which there is serious tension. When people are disagreeing about science and religion, it is usually about these issues.

That recognised, these disagreements are also arguably misdirected or overdone, and each chapter in part II, having surveyed the points of tension, also looks at how they can be and often are pacified. Time and again, it shows that where people are disagreeing, they needn't be. Time and again is not *all* the time, however, and because this book is ultimately more interested in outlining where people disagree than in persuading them that they shouldn't, each chapter concludes with a section on outstanding questions. It turns out that many of the disagreements regarding science and religion are misguided, or avoidable, or simply superfluous. But not necessarily all of them.

7

Metaphysics

Introduction

The question of what metaphysical stance is inherent in or necessary for the practice of science looms larger in American definitions than in British ones. 'Creation science' and Intelligent Design (ID), ever-present in debates over US education, are to blame. When the boundaries of science are being pressed from what is ultimately a metaphysical direction, those who wish to fortify them will naturally concentrate their energies at that point. Two of the five criteria in the Overton ruling mention the centrality of 'natural law' to science. The submissions to *Edwards v. Aguillard* stress how science is 'devoted' (a peculiarly religious-sounding word) 'to formulating and testing naturalistic explanations for natural phenomena'. And the judgement in *Kitzmiller v. Dover Area School District* emphasised science's 'rigorous attachment to "natural" explanations' *and* its 'rule ... by methodological naturalism'. By contrast, science's commitment to naturalism does not appear in either the short or the long version of the UK Science Council's 2006 definition.

There are, of course, plenty of conceptions of religion to which this apparent commitment to naturalism—the precise meaning of which we shall turn to presently—is in no way antagonistic. Those like *Malnak v. Yogi* in the US, which adopt Paul Tillich's idea of 'ultimate concerns', or *Hodkin* in the UK, in which Lord Toulson deliberately eschewed the word 'supernatural', are unlikely (or at least less likely) to find themselves in tension with science's apparent commitment to naturalism. Those that are informed primarily by Durkheim's concept of religion as a 'unified system of beliefs and practices', or by Geertz's as a 'system of symbols', or by Smart's practical, social, and material dimensions of religion are even less likely to find metaphysics a point of conflict. In this vein, as one science writer said to us, 'It feels to me as though nothing that is said in the New Testament that need affect the way we understand it and lead our lives today requires a literal belief in miracles' (#10). As we emphasise throughout this book, the conflict or concord between religion and science is predicated on the antecedent conceptions of each.

In this instance, however, the overwhelming evidence, from legal, philosophical, and sociological conceptions of science and of religion, not least from the expert interviewees to whom we spoke, was that there is a genuine point of overlap at this juncture. In spite of the claim of one philosopher that science has effectively answered all metaphysical questions,[1] the broad consensus was that (1) science is indeed grounded on certain metaphysical commitments that it cannot establish on its own, and (2) that religion is, for most people, inconceivable without its own (non-naturalistic) metaphysical beliefs. 'Religion without supernaturalism, I wouldn't understand what it would be ... I am not sure what we would have left' (#39). This is an area where the magisteria overlap.

Two points are worth noting here, before we move to a discussion of the potential for tension along this front. First, the overlap at this point does not necessarily result in discord. On the contrary, as we have already noted, there are some metaphysical commitments, like the apparent lawfulness of the universe and the ability of the human mind to grasp its truths, which are often cited as examples of metaphysical concord. Historically, the Judaeo-Christian concept of God as a lawgiver provided (some of) the raw metaphysical materials necessary for the scientific revolution, as Joseph Needham argued in his attempt to explain why Europe enjoyed a scientific revolution while (a technologically more sophisticated) China did not.[2] A number of philosophers contend that the same principle applies today.[3]

Second, some (similarly important) metaphysical commitments do not obviously come down on one side of this debate or the other. Beliefs, such as in the existence of external reality, of other minds, or of the reliability of the senses, may be legitimately judged to be 'basic'—in as far as they don't have 'wonderful epistemic warrant ... [but] we simply can't get away without assuming them' (#76)—without then being enlisted on the side of either 'conflict' or 'concord'. Overlap doesn't always mean taking sides.

These two caveats noted, there was a strong sense that the metaphysical overlap between science and religion was one in which the former's commitment to (some form of) naturalism came into non-negotiable conflict with the latter's commitment to (some form of) supernaturalism. This was more of a problem for some religions than others. Eastern religions, notably Buddhism (or, more precisely, certain formulations of Buddhism), were viewed as more readily in accord with the kind of naturalism required by science.[4] 'Buddhism is about understanding human psychology. It's

146 THE LANDSCAPES OF SCIENCE AND RELIGION

about understanding the world around you and recognising that things lack permanence, which science probably would agree with' (#99). That this applies to certain (often Western) secularised conceptions of Buddhism, such as those advocated by Owen Flanagan and Robert Wright,[5] rather than to Buddhism as it is practised by millions of its adherents worldwide, does not entirely blunt its impact on this debate.

Other religions, and in particular the Christianity that was the deep cultural background for most of the interviewees and served as their implicit template for religion, did have a problem with naturalism. Bluntly—and sometimes the opinion was expressed very bluntly—religion was predicated on a supernaturalism that was completely incompatible with science's commitment to naturalism. 'If you really want to believe in the supernatural, do. Personally, I think it's rubbish and I think that's one of the big reasons that makes science and religion incompatible' (#24).

This was just one of the more colourful examples of this point. Scientists and philosophers expressed astonishment that pre-modern beliefs in divine activity could persist in today's world. 'I find it very hard to understand the concept of god and the supernatural given that our context of the natural world is far, far broader than it was in the past and our natural understanding has been so successful' (#33). Historical examples of divine activity within holy texts were literally incredible. 'If you are believing in something that seems to contradict what we know about the laws of nature ... the literal truth of miracles in the New Testament, or the resurrection ... then there are problems' (#10). It was a view that drove some towards a more ethical, cultural, or ritual understanding of religion out of necessity. 'If you're expecting me to believe something because your explanation [contains] ... some supernatural elements ... that's not going to happen ... the smaller degree to which a religion does invoke a supernatural the more likely I am to be happy with it' (#9). Plenty of other examples could be added to these.

It is important to emphasise that this was more than mere contempt. People reasoned that the idea of the supernatural subverts the closed causality (or 'physical causal completeness') that is essential to the practice of science. Belief in any metaphysically independent, non-physical causes was incompatible with the principles of science, and risked opening the floodgates to any kind of 'explanation' that would undermine the entire scientific enterprise. Effectively, being able to deploy the wild card of 'God did it' rendered the entire game unplayable. As one respondent put it, and in the process neatly demonstrating the connection between metaphysical and social dimensions of science, the consequences of invoking the supernatural could be potentially apocalyptic, spelling nothing less than the end of human progress.

> As a species, we are unsurpassed in our ability to overcome environments of every single shape and space. We evolve in Africa, but we've conquered every continent and every life space, we've got space stations, and we're able to send people to Mars [sic]. We've done that only by treating the world as a naturalistic object. (#37)

It was no surprise, then, that some, in the manner of Lord Toulson, avoided the word supernatural altogether, on account of its being too 'emotive' (#91). Progress was founded on science and science was founded on naturalism and, therefore, supernaturalism, at least understood in this way, was a threat not only to science but also to the centuries of human development that had been constructed on science's naturalistic approach to the world.

Religion's commitment to supernaturalism

So far, then, so clear: in as far as science claims a necessary commitment to naturalism and religion to supernaturalism, there appears to be a genuine and possibly irresolvable metaphysical conflict in place. The complication comes in the nature and necessity of both of those commitments: whether these two options were the only ones on the table, and whether the dichotomy between them is as definitive as some clearly thought.[6]

In the first instance—we will turn to science's commitment to naturalism in the following section—the religious commitment to supernaturalism could be somewhat more hesitant and qualified than the blunt zero-sum relationship imagined. Those religious interviewees who voiced an opinion on the subject preferred to distance themselves from the category of the supernatural altogether.[7] This was not, to be clear, a renunciation of any and all metaphysical commitments for religion in favour of a solely ritual or 'ultimate concern' understanding. Those who cast doubt on the nature and usefulness of the 'supernatural' as a category were not simply embracing a narrowly cultural or ethical understanding of religion as the alternative.

Rather, they argued against the conception of the supernatural as a causal alternative to the natural, as if we were compelled to choose between 'God did it' and 'nature (whether in form of molecules, genes, brains, or humans, etc.) did it'. The idea that the 'spiritual' (the interviewee in this case preferred

148   THE LANDSCAPES OF SCIENCE AND RELIGION

to avoid the word 'supernatural' altogether) was 'something that is outside the natural that breaks into it' wasn't attractive or persuasive (#85). Nor indeed was it, at least according to one scientist/theologian, the way in which divine action had been historically understood.

> The laws of nature define the boundary between nature and supernature [and] if God wants to do something in the world, God has to break down the barriers that have been set up. Now, it is not clear to me that prior to this scientific revolution that people lived with this kind of really heavily dualistic view of the world. (#88)[8]

By this reckoning, the spiritual was not another thing set in contrast to the material, but rather an aspect of the material. In the words of one philosopher:

> Most sensible theologians would reject that dualistic framework and they would want to say that the supernatural is not reference to a second, super-natural, realm, which is problematically related to this [natural] realm [but] it's rather a dimension of nature. (#93)

In the same way that the same human action, for example coming to the aid of someone who has mobility problems, can be described and under-stood in several ways, e.g. in material terms (e.g. cognitive or muscular), in moral terms (e.g. showing kindness), or in social terms (e.g. obeying social norms), so might one event in the world be understood as both 'natural' and 'supernatural'. One theologian invoked Aquinas in her explanation of this.

> If we look at something like neo-Thomist definitions of divine action, you have God working in and with secondary causes. So, you could look at that through a scientific lens and see every cause and effect that you would expect, but not see any divine cause. [However,] a full description of the action would include the divine cause but not as something that is the action plus God, right? (#85)

This did not, of course, settle this conflict over the nature of the religious commitment to the supernatural, let alone the vexed question of divine action, and arguably it risked opening up a new and rather painful ethical front.[9] But it did at least offer some nuance on the potential natural–supernatural boundary dispute.

METAPHYSICS   149

That dispute was further nuanced by the observation made by a (non-religious) sociologist that there was nothing in principle antagonistic between science's naturalism and religion's belief in the *mere existence* of the transcendent or supernatural.[10] In line with the *Edwards v. Aguillard* submission, this kind of belief was simply beyond science's remit: a naturalistic discipline like science was not capable of adjudicating on the existence of non-natural phenomena or dimensions.[11] In the words of one interviewee, 'if you're fishing in the sea with a net that's two inches wide in the holes and all you ever catch is fish bigger than two inches, you'd be silly to say there's nothing in the ocean smaller than two inches'.[12] In this way, method and metaphysics were related.

The 'slippery slope' objection to this point, sometimes put forward by stricter adherents to naturalism, was that even such mere belief would, if followed to its logical conclusion, pose a threat to scientific explanation. If you start believing in God, you end up invoking God as an independent causal agent within nature, thereby undermining the entire scientific edifice. This was, however, at least in this interviewee's opinion, not a persuasive argument.[13] In effect, he argued, it was intellectually credible to believe in God without believing in an interventionist, still less a hyperactive, God.

Furthermore, as the same sociologist argued, there is nothing inherently illogical (still less dangerous) in engaging in the various traditional religious practices vis-à-vis this God, higher power, or transcendent dimension of reality.

> It can make sense to do things like pray and worship ... simply trying to acknowledge the existence of something else that might be out there and indeed think about various normative things to do with our responsibilities to each other, to the world and so on and so forth. (#28)

Prayer, worship, adoration, reverence, meditation, contemplation, or veneration of the 'supernatural' (though 'transcendent' would be a better word here), all as a means of morally and existentially orienting and inspiring the person, presented no necessary conflict with naturalism.

The real problem came with the conviction that this supernatural dimension (or entity or being) was actively involved with the natural world or could be successfully entreated to intervene in it.[14] This did appear to constitute a violation of the physical causal completeness that was foundational to the naturalism on which science depended. Honouring God was one thing.

150 THE LANDSCAPES OF SCIENCE AND RELIGION

Imagining that God interrupted the closed causality of nature to do things asked of Him was quite another.

This intercessory dimension of religion is not particularly prominent in either British or American legal definitions, but it did emerge among the disambiguated family features of religion, and it is, without question, central to the practice of millions of religious believers worldwide. In short, however much the nature of the religious commitment to the supernatural was clarified and qualified, there remained potential disagreement pertaining to the practice of prayer, and even more so the belief in miracles.

Science's commitment to naturalism

If religion's commitment to supernaturalism was complicated, so was science's to naturalism.

Numerous interviewees pointed out that renouncing supernaturalism in favour of naturalism was easy and obvious but resolved little. 'Many [of my fellow] secular humanists think, "Great! We have parked up supernaturalism, and the obvious place to go is naturalism" ... [but] to me, scientism—which is what that is—is at least as great a threat as some of the more fundamental religious beliefs' (#12).[15] Just because supernaturalism (or substance dualism) was not a satisfactory metaphysical home for science, it didn't necessarily follow that naturalism was.

The complexity began, as the above quotation intimates, with the variety of terms used. 'Naturalism', 'materialism', and, less often, 'physicalism' and 'scientism' were all deployed, often interchangeably, to describe the necessary metaphysical underpinning of science. Differences between them were drawn faintly, if at all.

'Materialism' was often employed but also often critiqued for assuming a rather crude, outmoded, and unconvincing 'Newtonian billiard balls' understanding of reality (#64). Matter, according to this view, was not what it used to be, and the spaces and forces that made up the once-solid atoms had pulled the rug from beneath old-school materialism.

Physicalism was occasionally used as an alternative, although this was not on account of any 'principled difference' between the two so much as simply because materialism had become 'associated with a rather vulgar account of matter' (#55).[16] Physicalism was materialism with better PR—'physical' being less vulnerable to reproach than 'material'.

Scientism, properly speaking, is a methodological rather than metaphysical term, referring to the belief that the scientific method is necessary and sufficient to describe and understand reality. However, it was still sometimes deployed, as above, as a cypher for naturalism, as the common but loosely used phrase 'scientific naturalism' intimates. Like materialism, scientism was rarely used neutrally.[17]

Naturalism was the preferred and usual term in this conversation, presumably because it offered the sharpest distinction to (religious) supernaturalism. It was, however, beset with problems around both its meaning and its relationship to science.[18] Attempts to define naturalism exactly floundered. 'Nature' and 'natural' were distinctly slippery terms, and our understanding of them has changed enormously over time, as our capacity to observe, manipulate, and alter reality developed. 'Two hundred years ago, radio waves, digital electronics, was all supernatural, right?' (#27).[19] The 'natural' was as contingent and 'constructed' a category as any other.[20] It could not satisfactorily be equated with the physical, or at least to do so risked excluding a number of scientific disciplines. It could certainly not be equated with the observable.[21] It could not be equated with the material, for the reasons against materialism outlined above. And even the (usually implicit) idea that naturalism meant closed physical causality came under scrutiny and stress through quantum theory.[22]

Numerous interviewees, particularly philosophers, attempted to finesse the definition by distinguishing methodological naturalism from ontological naturalism. By this reckoning, the former assumed closed physical causality for the sake of science, whereas the latter extended that assumption to reality, in the process abandoning belief in all non-physical entities, including God. 'I draw a fairly firm line between methodological and ontological naturalism. Science ... can't demonstrate the truth of ontological naturalism because that's a metaphysical position' (#97).[23] Others finessed naturalism in different ways, such as by drawing a further distinction within methodological naturalism, between 'essential' and 'pragmatic methodological naturalism' (#46), or by drawing a different one altogether, such as between 'conservative and liberal naturalism'.[24]

Although the distinction between methodological and ontological naturalism was a common one, it was not universally applauded. Some rejected it on the basis that intellectual honesty demanded that if we accepted one, we should accept both, i.e. if science demanded methodological naturalism (which it did) then we could safely assume the truth of ontological naturalism.[25] Others took almost the opposite approach, arguing that

152 THE LANDSCAPES OF SCIENCE AND RELIGION

methodological naturalism was itself an unwarranted and unsatisfactory assumption, especially for theists, and that we should therefore reject onto-logical and methodological naturalism altogether.[26]

In short, there was little consensus on what naturalism meant or entailed, and this uncertainty fed into the second problem, namely the extent to which it did indeed provide a satisfactory metaphysical basis for science. As with the wider question with which this book deals, many of the answers to this question turned on antecedent definitions. If you defined naturalism in a certain way—especially if you gave it a capacious or slightly vague definition or, better still, insisted that the only choice was between it and intervention-ist supernaturalism—then its status as a secure foundation was self-evident. For many, however, and by no means only those who might have had a prior religious inclination against naturalism, there were areas of science[27] that did not sit comfortably under (certain conceptions of) naturalism.

Examples were mainly drawn from the outer reaches of the physical sciences: 'eleven-dimensional theories' (#50), 'infinities and singularities' (#40), 'scientific laws' (#67),[28] 'dark matter and dark energy' (#40), 'hypothe-sising infinite universes [to which] we have no access' (#14), 'entities [within particle physics] which are not physical, but [which might] leave some sort of physical trace' (#81). In the words of one interviewee, 'as a theoretical physi-cist ... the things I studied way back in my PhD, are things that don't exist and that can't exist ... types of universe which we know are not our universe but you can still investigate them mathematically' (#81). Defining natural-ism in terms of nature, matter, the physical, the measurable, the observable, the existent, or even in terms of closed causal systems could make it difficult to accommodate these kinds of scientific ideas and disciplines. The result was a kind of naturalism that, in certain circumstances, was straining at the seams.

> I think that some philosophies of science are trying to reach towards the transcendent ...because they believe that materialism, or naturalism, is no longer sufficient ... some scientists are beginning to become open to this idea that pure naturalism, especially in its more materialistic variety, is no longer adequate, even within scientific fields. (#82)

In addition to this, as several philosophers pointed out, the problems with naturalism extended beyond the sciences. One (atheist) philosopher was eloquent in summarising naturalism's difficulties in terms of four Ms, namely 'morals, modal necessity, maths, and minds' (#42). Naturalism struggled to accommodate moral realism, the idea that moral beliefs are ultimately about facts concerning the world and not simply opinions thereon. It struggled with modal logic, the reasoning and weight behind terms like 'necessity' and 'possibility'. It failed to account for the truths of mathematics, turning many mathematicians instead towards mathematical Platonism.[29] And, it couldn't cope with the reality of consciousness in the human mind.[30]

All these points are, of course, highly contentious, and we will return to the question of minds in our chapter on anthropology, but it is worth briefly dwelling on this topic here, in the context of metaphysics, as it resulted in one of the genuine surprises of this research, namely the frequency with which panpsychism was mentioned and advocated.[31]

Panpsychism is the belief that consciousness—or mindedness, or mentality—exists 'all the way down' in reality, and that the consciousness that you and I experience is simply the aggregation of consciousness that already exists in our constituent material parts. A surprising number of interviewees spoke, without prompting, of panpsychism as a live alternative to naturalism or physicalism, an alternative that did not incur any kind of (untenable) dualism.[32] By their reckoning, 'consciousness is just an extra property of matter' (#17), 'fundamental to the universe rather than incidental' (#32). Matter is not inert but has an extra property, which corresponds, at our level, with consciousness. The first-person perspective is not generated by the brain so much as present incipiently in the material from which the brain is constructed. Importantly, panpsychism was not a religious alternative to naturalism. Interviewees were explicit about this. 'A lot of my fellow panpsychists are complete atheist secularists' (#17). But it was an alternative to the understanding of reality (that many believed to be) intrinsic to naturalism, namely a 'dead, essentially ... inert backdrop of particles and forces' (#36).[33]

The discussions around panpsychism serve as an example of how the supposedly hard and fast boundary between naturalism and supernaturalism could be blurrier than assumed. The naturalism underpinning science and the supernaturalism underpinning religion could clearly be understood in an absolute and non-negotiable way, but that depended on defining both in a certain way. 'If nature is mechanical unconscious and mechanistic then the only way God, angels or spirits can exist is supernatural' (#41).[34] More broadly, if naturalism was *de facto* defined as that which we understood and supernaturalism as that which we didn't, then a zero-sum relationship between them was inevitable.

154 THE LANDSCAPES OF SCIENCE AND RELIGION

> I've got a philosophical problem whenever people talk about naturalistic versus non-naturalistic explanations because when we find something utterly new and strange, such as Newton [did] ... then he just simply really expanded the definition of what's regarded as natural. (#64)

This was a remarkably common view, with numerous philosophers and scientists effectively understanding 'natural' as a cipher for 'known' and 'supernatural' for 'unknown'.[35] 'There is a sliding scale between naturalist and supernaturalist. As we learn more, we go down that scale and we take what was supernatural here, and we push it that way' (#27).

If, however, you have a more expansive definition (of both), the line between them begins to blur and the conflict is *to some extent* defused. 'If nature includes consciousness then naturalism is very different' (#41). One philosopher, for example, cited John McDowell, David Wiggins, James Griffin, and herself among those who had been inspired by Iris Murdoch to claim a form of naturalism in which meaning and values and 'all of the things that are central to what we're receptive to as human beings' are genuine and real and immune to physicalist reduction.[36] Mind, morals, meaning, and mathematics nibble away at the hard boundary of naturalism, as do quantum mechanics, string theory, multiple dimensions, and multiverses. In a similar way, the theological inclination to see the supernatural not as something external to and invasive of the natural, but as an aspect of or dimension within the (generously defined) natural, undermines the hard border of supernaturalism. Conceived this way, the metaphysical tension between science and religion can be negotiated—but not resolved.

The microcosm of miracles

The reason for this is only partly because some people, preferring their metaphysical commitments plain and undiluted, find any such finessing to be little more than evasion and equivocation. More substantively, it is because of a familiar topic of the science and religion debate.

Ultimately, however generously one defines naturalism, science still gravitates towards 'physical causal completeness' as, at the very least, a necessary methodological presupposition. At the same time, however subtly one defines supernaturalism, religion still gravitates towards belief in an entity (to use as vague a word as possible) that is beyond the immanent, 'closed', physical world. And while we have noted how the mere existence of these two 'systems' need not be a point of contention, some form of commerce between them is envisaged in the doctrine and practice of many religious adherents and, *in extremis*, that commerce is understood as not only 'God' breaking into the 'physical causal completeness' but also working *against* its processes and laws. In short, we come back to the question of miracles.

To be clear, miracles were at no point described as an essential part of religion in any of the legal or anthropological definitions we explored in part I. Nor were they mentioned amongst the disambiguated family features of religion generated by our research. Indeed, arguably it is only because Christianity forms the (implicit) template for what religion is for so many Western people that miracles get mentioned at all in this debate. And yet they do get mentioned. Like it or not, and even if they are ultimately simply a subset of a wider metaphysical tension, miracles are one of the things that people disagree about when they disagree about science and religion.

Understood in one way, miracles were the archetypal example of precisely the kind of supernaturalism that was deemed incompatible with science's commitment to naturalism. Science is predicated on the idea that natural (or physical or material) causes generate natural (etc.) effects. In as far as religion espouses non-natural causes for natural events, it risks undermining, and so conflicting with, the whole practice of science.

When those direct interventions into the natural order go with the grain of nature, they may be accommodated by means of a particular understanding of primary and secondary causality (though, as observed, this is not without its challenges). But when they go against that grain, there is undoubtedly a problem brewing. Miracles are where we see the metaphysical tension between science and religion crystallise.

This, at least, was the bluntest view of this issue. One philosopher protested that miracles had no explanatory weight and would not be accepted in any academic field and pondered why were they tolerated in religious studies.

> You had people at the time [of the Spanish Armada] saying that that was God blowing the invading Spanish ships offshore. No serious historian of the Elizabethan age today would entertain the notion that God might have blown those ships off course. So, why should we, as biblical scholars, or historians of religion, why should we entertain the notion that a supernatural explanation is the right explanation for something like the miracles of Jesus? (#61)[37]

156   THE LANDSCAPES OF SCIENCE AND RELIGION

Less antagonistically, some interviewees, both religious and non-religious, were willing to countenance the *possibility* of miracles even within a scientific worldview. For example, from a non-religious perspective, one said:

> I can imagine a religious scientist saying, I think this would actually be a consistent worldview, "I believe that the vast majority of things in our world are governed by oodles of science and all the natural processes that we know of. I also believe that there's another level of reality that's non-naturalistic, and every so often, somehow or other, that other plane breaks into our natural one, does something and then sort of disappears again." (#28)[38]

That acknowledged, the same interviewee immediately went on to say that however much this might be a *possible* worldview, 'it's not a view that I think many people would find very attractive' (#28). Although he didn't draw the analogy, the approach seems to be a variation on ID, with miracles being invoked to explain things that weren't otherwise explicable by natural causes.

The problem with this approach was not simply that it was inelegant and unattractive. It opened the door to a number of deeper problems with the concept of miracles themselves. First, the idea of miraculous divine intervention afforded 'God' some interventionist credentials but only at the expense of hanging him on a rather painful ethical hook. One chemist acknowledged that 'I think that it's logically possible to say that ... the universe obeyed certain laws except when God decides not to' (#46). However, he went on to say:

> I don't quite understand why religious people are comfortable with that point of view because if they allow God to intervene miraculously, then surely he is ... morally responsible for all the things he allowed to happen by not intervening. (#46)

In much the same way as eliding natural and supernatural risked making God (or a higher power, etc.) responsible for a whole load of nasty things in the world, so insisting on the possibility of occasional, interventionist, against-the-grain miracles risked making God responsible for a whole load of nasty things to which God appeared indifferent by not intervening when He presumably could. Solving a metaphysical problem simply led to an ethical one.

Second, there was the problem of the gap between the possibility of miracles and their reality. Opinion was divided on whether the miraculous was truly investigable. On the one hand, the uniqueness of miracles was judged to put them beyond the purview of science.[39] On the other, some miraculous claims were not 'unique'—those concerning the efficacy of prayer for healing, for example—and, that being so, even if the interventions were unobservable in themselves, they should be detectable in the data.[40]

Those interviewees who referenced large-scale studies on the efficacy of intercessory prayer pointed out that their conclusion was usually negative. There was nothing to be seen here.[41] In addition, those who had themselves in anyway been involved with miraculous, paranormal, or parapsychological investigations reported that they always drew blanks. 'Some of the work I did in India was looking at the Gulf people of India who claim to be able to materialise objects and so on. I would be pretty sceptical about the evidential base. I don't think there is much going on there other than trickery and self-delusion' (#25).[42] In short, however theoretically possible the miraculous might be, empirically it didn't get off the starting blocks.

Even if it had, however, and, for example, evidence for the positive effect of intercessory prayer on patients had been detected,[43] there remained a third problem, this time with the very concept of the miraculous. The same psychologist who had investigated the topic put it this way:

> People constantly come up with claims that are supernatural. We constantly test them. If those claims turned out to be true, you would alter what you believe about the world in order to incorporate that new evidence. (#25)

In effect, as soon as something that is claimed as super- or non-natural is confirmed to have occurred, it would no longer have a claim to be super- or non-natural. If it happened in nature, it would fall, however peripherally, into the category of 'natural', albeit 'natural and highly unusual'. Put another way, if a miracle is 'by definition ... something ... that breaches the scientific rules ... or laws' (#28), then for something truly to be a miracle, those drawing that conclusion must have a complete and comprehensive understanding of these rules and laws, and so an equally comprehensive understanding of what is and isn't possible. If they don't, and few claim to, then there is a good chance that what is claimed to be miraculous is, in fact, merely a rare and still-less-than-fully-understood part of nature.

158 THE LANDSCAPES OF SCIENCE AND RELIGION

One biologist made the same point regarding Big Miracles, so to speak, meaning miraculous proofs of God. 'In a trivial way if God was to appear—whichever god you want—and say "I've had enough of this, I'm going to clear it up once and for all, here I am". And appeared in whatever form, in whatever domain, immediately at that point, he becomes a scientific entity' (#35). And (presumably) being a scientific entity—meaning an entity amenable to the investigative methods of science—'he' would cease to be God.

Approached this way, the very concept of a miracle is dubious. The best it can hope for is to remain inexplicable, but this is not so much confirmation of the supernatural as of human ignorance. 'A miracle is basically something that you can't explain. So, even if it is technically explainable, we can't explain it yet' (#38). By this reckoning, the miraculous is really an epistemological feature rather than a metaphysical one, effectively meaning 'that which we don't yet understand', in much the same way, as we saw above, that supernatural is often idiomatically a cipher for 'unknown'.

This might be seen as deflating all religious claims on the miraculous in a rather unceremonious manner, and there were religious interviewees who resisted any such negotiation of the miraculous, just as there are obviously many millions of religious believers who adhere to a straightforward occasional, interventionist, against-the-grain concept of miracles. But it is worth mentioning, in passing, that such a renegotiated concept of miraculous—away from the straightforward transgression of the laws of nature—is not necessarily at odds with the New Testament concept of miracles.[44]

There was, of course, no established idea of the laws of nature in the ancient Near East, and the Greek New Testament has no word 'miracle', which derives from the Latin word *miraculum*.[45] The New Testament words that modern translators render as 'miracle' are δύναμις (*dunamis*) meaning 'power', 'might', or 'strength'; τέρας (*teras*) meaning 'wonder' or 'marvel'; and σημεῖον (*sémeion*) meaning a 'sign'. In the New Testament, *dunamis* is used, in its various forms, 120 times, *teras* 16 times, and *sémeion* 77 times, and none is used exclusively of events that might have a claim to be a transgression of natural laws.[46]

That does not, of course, mean that they do *not* refer or cannot be used to describe 'transgressive' (so to speak) miracles. The very fact that they are used to describe actions of Jesus—healing, feeding, controlling nature, etc.—that clearly exceed natural human powers strongly suggests that the words were being used to describe something that was perceived to be very close to our modern, popular conception of the transgressive miracle. However, it is nonetheless worth noting that (1) hard and fast laws of nature were not in the mental universe of those who described these events and (2) the words they use tend to focus on the remarkable, wonder-inducing, *significance* of the events for those who witnessed them, rather than attempting to define their metaphysical status.

None of the interviewees made this point, perhaps because only a tiny minority were biblical scholars or knew any New Testament Greek. But a number did gesture in this direction, in two ways. First, they pointed out the word miracle had a powerful idiomatic meaning today that was arguably close to the way it was used in biblical texts. One, citing Colin Humphreys, former Professor of Materials Science at the University of Cambridge and 'a very observant Christian', claimed that 'we've got it all completely wrong about what miracles are ... because ... people today when something unusual happens will say, it's a miracle, I've won the lottery ... maybe people were saying that in the New Testament' (#20). Second, they preferred other terms to miracle, terms that lacked the latter's blunt and uncompromising metaphysical resonance. 'I'm not quite sure I'd use the word miracles. I want to use the word mystery ... [or] something transcendental' (#57).[47]

There were, and will be, plenty of people, both non-religious and religious, who find this equivocation around the meaning of 'miracle' no more persuasive that the negotiation of the boundary dispute between natural and supernatural. We are not, however, seeking to endorse either position. Rather, we are trying to clarify what it is people are disagreeing about when they disagree about science and religion. One area of that disagreement is metaphysical, and within the metaphysical part of the landscape, wrangling over the possibility, actuality, and very meaning of the miraculous is an important local skirmish. For some, it is an entirely justified brawl—miracles symbolising a much wider and more profound stand-off between the metaphysical stance of science and that of religion. For others, it is a mere scuffle, and a largely baseless one at that.

Outstanding questions

There is without doubt a perception that not only do science and religion both have a metaphysical dimension to them—in the terms we have been using, the family definition of each has a metaphysical

160 THE LANDSCAPES OF SCIENCE AND RELIGION

feature—but also these metaphysical commitments are comparable; they engage or 'overlap', so to speak. When that engagement is framed in terms of naturalism and supernaturalism, or is focused on the meaning and nature of the miraculous, it amounts to an outright dispute. It is clearly one of the things that people genuinely and legitimately disagree about.

However, what emerges from careful exploration of this topic is that although it is quite possible to construe this dispute in absolute and non-negotiable terms, there are more nuanced understandings to which many, and certainly not only religious thinkers, are alert. In effect, just as the overarching question about science and religion fragments into sub-questions about metaphysics, epistemology, anthropology, and the like, so those questions fragment into further issues, which demand the attention of anyone who is serious about the science and religion debate.

In the case of metaphysics, for example, once the blunt opposition has been defused, various sub-questions remain. Is there a concept of the non-natural (to avoid using the 'emotive' word supernatural) to which religion is wedded that doesn't simply involve the invasion and subversion of the physical causal completeness which seems to be foundational to the naturalism on which science depends? Does methodological naturalism demand ontological naturalism, and, conversely, is ontological naturalism capable of explaining or sustaining mathematics, moral reasoning, conscious minds, and the like? Does the much-discussed (if not established) idea that some (arguably non-natural) truths, for example about minds, morals, and mathematics, are accessible to intuition have anything to say to longstanding religious claims that some individuals have intuitive access to (non-natural) spiritual truths?[48] How far does the concept of the miraculous demand not only commerce between but also invasion of the natural by the non-natural? Or, alternatively, how much is the category of the miraculous actually epistemological ('the miraculous as the mysterious') or anthropological ('the miraculous as the wonder-inducing or personally significant') rather than metaphysical?

These are complex questions, not amenable to straightforward answers, and easily dismissed as equivocation by those already heavily invested in conflict (or indeed harmony) between science and religion. But if we are prepared to focus down within the wider landscapes of science and religion, and hone in on the details, we will find that this is where the metaphysical disagreements or, more positively, debates really lie.

Notes

1. 'An awful lot of the classic metaphysical questions, and that includes the ontological questions, have largely been answered by science.' (#14)
2. 'It would be difficult to overestimate the effect on all occidental thinking of the Christian era of these ideas in the Hebrew scriptures—"The Lord gave his decree to the sea, that the waters should not pass his commandment"—which provided a powerful model for lawfulness.' Needham, Joseph, 'Human laws and laws of nature in China and the West (I)', *Journal of the History of Ideas* (1951), 12(1), 22.
3. 'We believe in natural laws and I don't think it makes any sense to believe in laws unless you believe in a law-giver.' (#83)
4. We are trying to avoid the pithier phrase 'scientific naturalism' in this discussion as this is commonly used to describe not only a form of naturalism but also one that is comprehensible through the scientific method, thereby eliding two different factors (metaphysics and method) that we have elected to keep separate.
5. Flanagan, Owen, *The Bodhisattva's Brain: Buddhism Naturalized* (Cambridge, MA, 2013); Wright, Robert, *Why Buddhism Is True: The Science and Philosophy of Meditation and Enlightenment* (London, 2018).
6. Cf. the words of one philosopher, who was, interestingly, one of the more anti-religious interviewees: 'I think one of the problems in religion, and one of the problems in science, and one of the problems in the debate between the two, is the use of these binaries' (#61).
7. 'I do think it's important that religious discourse gets away from the supernatural. It's almost a kind of marker of territory.' (#84) 'I'm wary of using words like supernatural because those come loaded with a whole history that I don't necessarily want to invoke all that they mean.' (#85)
8. This is arguably true, at least historically speaking, but it does invite another question, not explored here, namely whether a commitment to the lawfulness of creation (which, as we have seen, was judged to be rooted in Judaeo-Christian theology) therefore (ultimately? inadvertently?) does necessitate a natural–supernatural dualism.
9. Namely, God becomes a primary agent involved in much of the secondary pain and evil in the world.
10. 'I think that a belief in a supreme being or a higher power or a creation or whatever, is a respectable metaphysical belief. It's not one I happen to share, but I don't think that there's anything wrong with it as a belief or as part of a world view.' (#28)
11. Science is 'simply not equipped to evaluate' such claims, according to *Edwards v. Aguillard*.
12. 'By limiting what we are looking for, I don't think it at all follows there can't be more than that. It's just saying we don't have the equipment or the ability to say anything about that. I think at best it could be agnostic about ontological naturalism.' (#85) The comments were made on the distinction between methodological and ontological naturalism, to which we shall turn.
13. 'The objection of somebody like Dawkins would be once you accept that idea then it's very hard not to, by way of some slippery slope, get into the idea that this supreme being is interfering in various ways in natural processes and so you kind of automatically arrive at a position of conflict between the metaphysics and the scientific perspective. I'm not so sure that's true.' (#28)
14. 'There are lots of sensible things that might be going on when one is being religious that don't necessarily have to do with imagining interaction between that natural world and the non-natural one.' (#28)
15. In a similar vein, 'I would certainly agree that science has to be firmly naturalist [but] you have to get into some questions as to what the hell you mean by that' (#40).
16. 'Philosophers for some reason started talking about physicalism rather than materialism because I think they thought it sounded less nasty, less reductionist, but I don't think there's a principled difference between the two terms.' (#55)

162   THE LANDSCAPES OF SCIENCE AND RELIGION

17. For example, 'scientism is used in that way, often with a slightly patronising twist as well. There's a sort of, "what an idiot, what an unsophisticated fool this man is [for believing in] scientism". (#42)

18. Tellingly, the *Stanford Encyclopedia of Philosophy* remarks early on in its lengthy entry on naturalism, 'It would be fruitless to try to adjudicate some official way of understanding the term ... The important thing is to articulate and assess the reasoning that has led philosophers in a generally naturalist direction, not to stipulate how far you need to travel along this path before you can count yourself as a paid-up "naturalist"' (https://plato.stanford.edu/entries/naturalism/).

19. 'I don't know what natural is because humans have fundamentally altered every aspect of this planet over the last 100 thousand years.' (#35) 'What was observable scientifically a hundred years ago looks very different to what is observable scientifically today.' (#79) 'Some of the things we regard as supernaturalism would have been deemed as an extension of science 100 years ago as well.' (#65)

20. 'Nature is a construct, nature is a particular way and a word that has many meanings.' (#90)

21. 'Natural gets equated with physical, [with] things that are tangible ... But there are a lot of things science hypothesises that aren't like that.' (#14)

22. Although, as some pointed out, quantum theory did nothing to challenge closed causality at a Newtonian level. It could be deployed by 'both sides' in rather question-begging ways. 'We talk about instabilities and quantum vacuum. And with a little bit of fiddling, you can get a sudden instability in quantum vacuum that gives you something for nothing. You can get a universe from nothing. [But] it is a rather posh nothing.' (#12)

23. In a similar vein, 'Here is a sort of methodological naturalistic starting point that's taken seriously by the scientist and that confines her enquiry to causal explanations ... within the natural world, in figuring out why things happen and explaining why they happened.' (#93)

24. 'Conservative naturalists think the only data for science is observation [or] experiment ... But there might be other data, [such as] the reality of feelings and experiences.' (#17)

25. 'Methodological naturalism [is] perceived to lead you into ontological naturalism ... [but] some of those reasons [for that] I suspect are as much sociological as anything [else].' (#97)

26. On whether it was unwarranted, one interviewee referenced Andrew Torrance's argument in his paper 'Should a Christian adopt methodological naturalism?', *Zygon* (2017), 52(3), 691–725, with which she disagreed. On whether it is unsatisfactory see comment from #94: 'If you investigate assuming naturalism, if Christianity is true, you are going to come to singularities. Not simply at the beginning of the universe, but maybe at the beginning of life, at the beginning of consciousness, and things like that. It arbitrarily cuts things off. It is saying, "This is what the universe is like, and I am going to apply it to science." So, it is an a priori philosophical commitment.' We will touch on some of these issues later in the chapter.

27. More precisely, following the discussion in part I, disciplines that were commonly understood as falling within the family definition of science.

28. 'What's a scientific law? What's a law? Show me a law, is it in nature? It's a metaphor for God's law. It's just a metaphor, it's not a natural thing.' (#67)

29. 'Many mathematicians are mathematical Platonists, and they think they're exploring a kind of non-natural realm of numbers and triangles and so on. They think it's really out there, mind independently this reality, but it's not empirically available to us.' (#42)

30. To this list, another philosopher added a possible fifth M, 'meaning', although she herself was not entirely persuaded by this. 'I think someone like John Cottingham thinks that unless you buy into the transcendent perspective or external model in that sense you've got to say that meaning is self-generated and I think I'd want to resist that on the grounds that it's possible for example to be a value realist, a realist about meaning, without having to be a theist' (#93).

METAPHYSICS   163

31. Before we embarked on this project, we did not anticipate panpsychism coming up at all in the interviews. Not only was it mentioned by quite a few interviewees, but also it was vigorously defended by at least four of them.

32. Confusingly, at least one of those who planted their flag firmly in the panpsychist camp justified their views as an alternative to scientism rather than naturalism or physicalism. It was another example of the elision of terms here.

33. For some, this apparent challenge to naturalism from consciousness could be extended to the question of the human as a whole. 'This is where naturalism ends, right? Our conception of ourselves is not a fully fledged natural phenomenon.' (#14) In so doing it moved onto the anthropological part of the landscape.

34. In a similar vein, regarding supernaturalism, 'the thing about supernaturalism is that everything depends upon, again, how you understand it' (#93). More generally, 'it's setting up a dichotomous relationship by defining things which may not be really reflective of reality' (#78).

35. 'I think that to call something supernatural, that's an artificial distinction because that's really just saying the known laws of science versus the ones that we don't know yet, or we haven't discovered yet.' (#45) 'I think naturalism has a habit of encroaching upon supernaturalism.' (#39) 'The progress of understanding of the world has been a progress of naturalising phenomena, fitting it into something which doesn't require appeal to those agencies I was talking about, something beyond the veil of nature that explains their causal origin.' (#61) 'It would be hard to imagine a science that couldn't incorporate entities that we now consider to be supernatural.' (#25)

36. '[We] wanted to question that very narrow way of thinking about the limits of nature. There is so much that it cannot accommodate, all the things that are important to us as meaning-seeking human beings who are interested in morality and receptive to values, etc. ... those naturalists who were inspired by Murdoch, what she called "the presupposition to naturalism", and they say that true naturalism, they haven't always put it in quite these Murdochian terms, but true naturalism has to be non-scientistic or anti-scientistic and it has to embrace all of the things that are central to what we're receptive to as human beings.' (#93)

37. In a similarly critical, though more ameliorative vein, one biblical scholar argued that miracles 'should be bracketed out because, could anyone really prove that God intervened in history and resurrected a human being?'. This was an argument for historical self-discipline based on metaphysical agnosticism rather than outright rejection. As he went on to say, 'Could a historian prove or show the likelihood of a group of people who believed this happened? Yes, I think you can do that. I just don't know how it's possible for a historian to go beyond that and say, "therefore, this is supernatural"' (#65).

38. From a religious perspective, 'the sciences can, to some degree, proceed as if miracles don't happen, or at least proceed as if the explanation to their question isn't going to be "God did it". Now occasionally God might do something, and I think a scientist can believe that occasionally God might do something, and they can even, in certain contexts, pray for it.' (#83)

39. 'I don't think science has disproved the existence of miracles because miracles are singular events and I don't know if there's anything falsifiable about the claim.' (#28) That said, one sociologist pointed out that the Catholic church has a whole department, the Dicastery for the Causes of Saints, dedicated to investigating the veracity of claimed miracles, which clearly assumed some capacity for falsification, if not actual verification.

40. Thus, most fully: 'Just because the supernatural is unobservable does not mean that it's empirically off-limits ... If you've got evidence that heart patients are recovering when people pray to God and ask him to intervene, and there's absolutely no way that we can see that that prayer would have an effect on the heart patients by any naturalistic mechanism, any kind of mechanism operating within the spatiotemporal universe, there just aren't the connections there to have that result, then you would have pretty good evidence that this was working in some way, this would be some evidence that there was some kind of

164 THE LANDSCAPES OF SCIENCE AND RELIGION

being or agent beyond the natural order that was, in fact, intervening in response to these kinds of requests' (#28).

41. 'When you start to drill down and run well-controlled studies, the effects in some of the messier studies fade away.' (#25) 'There have been two multimillion-dollar funded studies into the efficacy of petitionary prayer, one run by somebody who believes in the power of petitionary prayer, so they're not really biased against the idea, both of which found ... no evidence for petitionary prayer working on heart patients.' (#42)

42. The conclusion would then be predictably Humean. 'In every case of a miracle, every case of an alleged miracle which I've gone into it, it's more plausible, much more plausible that the account is garbled than that the miracle itself occurred.' (#46)

43. Note that this is the positive effect of petitionary prayer on patients as opposed to on people who pray. The positive effect of prayer on pray-ers themselves is well established now.

44. The miraculous does, of course, appear in many religious traditions around the world, and so what the New Testament has to say about miracles cannot be taken as representative of wider religious traditions. However, as has been observed, and as is illustrated by the examples of miracles cited by interviewees, the miracles of Jesus and more widely of the Christian Bible formed the template in this discussion.

45. Itself meaning 'object of wonder'.

46. For example, 'To one he gave five bags of gold, to another two bags, and to another one bag, each according to his ability [δύναμιν]' (Matt. 25.15). 'For false messiahs and false prophets will appear and perform signs and wonders [τέρατα] to deceive, if possible, even the elect.' (Mark 13.22) 'Many people, because they had heard that he had performed this sign [σημεῖον], went out to meet him.' (John 12.18)

47. In a similar vein, 'nobody will bat an eyelid if people say I had a lightbulb moment or I had a kind of epiphany not in the religious sense' (#91).

48. See, for example, Parsons, C., 'Platonism and mathematical intuition in Kurt Gödel's thought', *Bulletin of Symbolic Logic* (1995), 1(1), 44–74. More generally, 'Moral Non-Naturalism' in *The Stanford Encyclopedia of Philosophy* (Fall 2014 edition) (https://plato.stanford.edu/entries/moral-non-naturalism/).

8
Methodology

Introduction

In the absence of any sustained and orchestrated attack on science's metaphysical front in the UK, the Science Council's 2006 definition ignores any discussion of super/naturalism and focuses almost entirely on science's methodological and epistemological characteristics. Science is a 'systematic methodology based on evidence', according to the shorter definition, which is elaborated to highlight its commitment to evidence, observation, experiment, repetition, critical analysis, induction, and verification in the longer definition.

The same focus on method(s) was evident in our disambiguated 'family resemblance' definition of science, as well as in various US legal discussions, albeit that some of these bear the mark of the creationist/ID challenge even within the methodological parts of their discussions. The *Dover Area School District* submissions, for example, emphasised science's dependence on 'empirical, observable and ultimately testable data' but did so in deliberate contrast to 'the appeal to authority', 'revelation', or 'philosophical coherence' which are the apparent alternatives. In short, no matter how foundational a metaphysical stance is to science, in as far as the overall family definition of science has one dominant feature, it is methodological.

This is not the case with religion, for which the content of belief has as much salience as the exercise of belief itself; *what* religious people believe is as important as the fact they *believe* in it. That acknowledged, that exercise of belief is still important, as intimated by the fact that the word 'religion' is so commonly interchanged with 'belief' or 'faith'. Although the centrality of belief (as opposed to, say, ritual or community) is, as we have observed, hotly contested, it is nonetheless an important feature with the family definition of religion. Religion is not *only* about the reasons for and ways in which people believe, but it is partly about that.

Crucially, however, this partially significant feature within religion assumes much greater importance within the science and religion debate, precisely because method and epistemology are so central to the idea of

166   THE LANDSCAPES OF SCIENCE AND RELIGION

what science is. In as far as science is characterised by one feature, it is its method(s), and therefore any conversation it has with other human activities, like religion, naturally focuses on questions of methodology. As a result of this, we consider methodology to be a second major point of connection, and potential conflict. When people are disagreeing about science and religion, they are often disagreeing about the process by means of which each acquires and holds the beliefs that it deems to be true.

Faith as a methodological shortcut

Conflict is no more inherent within this methodological part of the science and religion landscapes than it is in the metaphysical one we explored in the previous chapter. Quite apart from the fact that, as one sociologist warned, not many people engage in the science and religion debate this way,[1] it is perfectly possible to trace methodological correspondence rather than conflict here. In the words of one anthropologist, 'science is simply a methodology ... a way of asking questions. In that sense, it's not different to theology [albeit] theology relies on different sources of evidence to inform its arguments and discussions' (#51).

Such an ameliorative approach to the methodological comparison was not the norm, however. That was better summarised by the analogy made by the same sociologist cited above who remarked that 'in religions, you can have trump cards. You can whip out "God says", or "God told me"'. Science, by comparison, 'does not permit trump cards' (#21). Science demands its disciples take a long, slow, hard, indistinct, and often-forking methodological route to the truth. By contrast, religions believe in—indeed, religious belief *is*—the shortcut.

This point could be, and often was, made in a sweeping and indiscriminate way. Religions champion 'blind faith' (#24). Faith is 'valuing belief despite evidence, or even because of the lack of evidence' (#9). It is an alternative to evidence.[2] It makes a virtue out of believing things that 'cannot be rationally justified' (#46). Doubting Thomas is the religious justification for not asking too many questions.[3] Tertullian's *credo quia absurdum*—'I believe because it is absurd'—epitomised this approach.[4] On occasion, the criticism slid from the sweeping to the straightforwardly polemical. 'I don't have to be an expert at fairies at the bottom of the garden to know that I think that's probably incompatible with science ... [it's] basically ... barbaric, bronze-age superstition' (#18).

The same points could be made in more targeted ways. Some religious believers made a virtue of the lack of evidence.[5] Some used the word faith 'as a kind of bolt hole' (#5). Dogmatism was real, and it was a real problem, but it was not limited to religious believers. Such criticisms were levelled by theologians as much as scientists, religious interviewees as much as non- and anti-religious ones.

More perspicacious still were those methodological criticisms that went beyond fingering dogmatic, ignorant, or incurious segments of the religious population, to pinpointing what it is in religion's approach to acquiring and holding beliefs that is in tension with science's methods. Broadly speaking, those who did locate tension here drew attention to five ways by means of which religious beliefs are formed which they judged in some way incompatible with the scientific approach. Religious belief depends on *revelation*, which is 'pretty firmly at odds to the appeal to scientific inquiry' (#5). It appeals to *tradition*, often 'accepted without challenge' (#72), which is antithetical to science. It relies on *authority*—'presumed authority ... unassailable authority' (#13)—which is fundamentally unscientific. It treats subjective *feelings*[6] or *experience*[7] as legitimate evidence. And it grounds itself on *holy* or *sacred texts* that are factually unreliable, or essentially mythical,[8] or treated with unwarranted authority,[9] the interpretation of which is inconsistent,[10] or unsystematic,[11] or just plain subjective.

Sometimes, these different approaches could make awkward bedfellows. Experience has great potential to undermine tradition, for example. However, they could just as easily reinforce one another, thereby cementing some already-unwarranted beliefs.

> The experience you have is obviously completely shaped by the religious culture in which you happen to live. It's terribly rare for a white Christian-culture Brit to have a Hindu religious revelatory experience ... on what grounds does one say that the Christian religious experience provides for real content, whereas the religious experience of Sufis and Hindus isn't? (#76; #3)

Once again, while these objections could be raised polemically,[12] usually they were not. Similarly, while some argued that this all amounted to religious people (or authorities) being fundamentally stupid, ignorant, incurious, and incapable of using reason, this was not the default position. Some anti-religious critics willingly recognised that religious arguments did marshal reason (they just didn't find them persuasive[13]), while others pointed

168 THE LANDSCAPES OF SCIENCE AND RELIGION

out that if there was a problem with irrationality here, it was not limited to religious folk. In the colourful words of one sociologist:

> Modern non-religious people are quite capable of believing the most dreadful bollocks. The alternative to a shared religion like Christianity isn't atheism, or at least it is not articulate humanistic atheism or rationality. It is freestyling nonsense, with every individual putting together their own particular package of complete bollocks. (#21)

At heart, then, the methodological criticism could be shrewd and sincere—in essence that the claims that are central to religious belief are based, acquired, or sifted in a way that is at odds with the way science works.

We have, in part I of the book, already touched on the way that science works—or was thought to work. However, it is worth revisiting this briefly at this point because it was so frequently compared and contrasted to religion's purportedly defective approach, and therefore it was foundational to the methodological conflict narrative of science and religion.

If religion relies on revelation, tradition, authority, feelings, experience, and the subjective interpretation of holy texts to derive its beliefs, science relies on close and repeated observation; experimentation and quantification where possible; critical analysis and induction to the best explanation; attempted falsification; and, ideally, prediction. More generally, but also more important than the specific constituent elements of science's method(s), is its general ethos or attitude, the idea that science's approach is inherently inquisitive—curious, sceptical, self-critical,[14] argumentative,[15] interrogative, restless—in contrast to the religious approach to belief and knowledge, which is not.

By this reckoning, science questions authority, religion relies on it.[16] Science overturns tradition, religion builds on it.[17] Science rewrites its textbooks, religion preserves its.[18] Science values epistemic humility, religion certainty.[19] Science is organised doubt, religion institutionalised faith. Scientific knowledge is provisional,[20] religious knowledge claims to be definitive. And, as a result of all this, science is cumulative and progressive, whereas religion is static and stagnant, and always needs to be kicked dragging and screaming into the modern world. The link to socio-political areas of tension, to which we will turn in a later chapter, will be obvious.

To be clear, and at the risk of repetition, this was not mere invective (or, rather, only rarely so). Interviewees cited examples of institutions and groups that illustrated this dichotomy: there really were religious believers and institutions whose approach to (their) belief(s) was incompatible with science's method(s) for the reasons stated above. They also acknowledged reservations and nuances within the dichotomy; the skirmish over methodology in this part of the science and religion landscapes was real, but it might not be inevitable.

Rather, this is simply a sharp crystallisation of one of the more contested areas within the wider science and religion debate. One of the things people disagree about when they disagree about science and religion is their perceived, fundamentally different approach to acquiring and holding beliefs.

Re-evaluating the methodological dichotomy

The way in which a scientist hypothesises, tests, replicates, and doubts their way to their eventual (tentative) beliefs is, then, not only different to, but fundamentally at odds with the way in which a religious believer assents and accepts their way to theirs. The contrast is even more pronounced when the scientist is working by the book, so to speak, punctiliously following the heavily theorised path of 'the' scientific method, while the religious figure finds themself believing as a result of a messy, compromised, inarticulate faith journey, shaped by upbringing, culture, social pressure, personal experience, pastoral support, personal hopes, and a passing acquaintance with their new religion's holy text. The comparison is obvious.

But it is also unfair. One science journalist inadvertently alighted upon why with a telling metaphor. 'Religion sees itself as being the quest for the truth, whereas science, in its strictest sense and *in its Sunday best* as it were, is the quest for provisional understanding which may be revised later' (#13; emphases added). The interviewee here was not questioning the fundamental methodological dichotomy outlined above. Scientific and religious approaches to the development of beliefs were, in his opinion, different and arguably incompatible. But the science that was different to religion here was science-in-theory, science in 'its strictest sense', or, more colourfully, 'in its Sunday best'. Comparing the scientist who conscientiously experiments and infers their way to provisional beliefs with the believer who obediently assents their way to The Truth is like contrasting science in its Sunday best with religion in its dirty, weekday overalls. It is effectively comparing a normative conception of one with a descriptive conception of the other, science-as-it-should-be with religion-as-it-is. And it is when

170   THE LANDSCAPES OF SCIENCE AND RELIGION

the skewed nature of this comparison is acknowledged that the apparent methodological conflict within the science and religion landscapes becomes more honest, more interesting, and perhaps more tractable.

In the following sub-section, we will re-examine the assumptions behind the idea that the scientific approach to acquiring knowledge is fundamentally in tension with comparable religious approaches. In so doing, we'll show that, according to many of the practising scientists, not to mention the sociologists and philosophers of science, to whom we spoke (even those wedded to the idea of a methodological dichotomy), some elements of scientific knowledge acquisition, such as intuition, experience, subjective judgement, the need for written authority, institutional recognition, and even 'faith', veer perilously close to the religious approaches against which science is pitted. In effect, the dichotomy may not be as severe as is claimed. In the sub-section that follows on, we will turn to the religious side of this particular conversation, and explore the same territory but from the opposite side.

Revisiting scientific methodology

The relevant discussion in part I noted that the actual practice of science was a good deal messier than the idea of 'the scientific method' suggested. The source of this messiness is two-fold: (1) that the theorised method(s) of science are themselves more varied, complex, and dependent on imagination and subjective judgement than sometimes claimed; and (2) that the actual practice of science often falls short of the theory and introduces a great deal of fallibility into the process. We will deal with these in turn.

The messy methods of science

Messy need not mean defective or sub-standard. Indeed, messy can be thoroughly positive. Science demands the exercise of imagination. 'You can't be an astronomer without imagination. You can't really be any kind of scientist without imagination' (#87). It requires a degree of creativity. It sometimes depends on intuition, even on moments of inspiration. Whatever else, it is more than the repeated and undeviating application of an established method. It requires originality and resourcefulness.

Messy can also simply mean compound, multifarious, and in need of discriminating application. The employment of science's various methodological steps—observation, experiment, repetition, critical analysis, induction, etc.—demands more than their formulaic implementation. They require vision, resourcefulness, discernment, and judgement. Conceiving and planning an experiment can require imagination and foresight. Science, we were told, is 'full of human interpretation' (#74). Analysing results may call for shrewd or intuitive judgement between different 'theoretical virtues' such as 'simplicity, elegance, intelligibility' (#17). Beauty is an important, if contestable, heuristic.[21] Moreover, interpretation and judgement are learned by doing, the practical, hands-on, 'lived' experience of being a scientist that goes beyond theorising. Indeed, the whole thing can be equated with a kind of apprenticeship. In the words of one cosmologist:

> A lot of philosophers of science actually don't understand science, and what I mean by that is that science is a complex messy business which is learnt by apprenticeship rather than by systematic formula teaching. So, science is an art not a science, if you see what I mean. (#16)

In addition to imagination, subjective judgement, and patient apprenticeship, science requires respectful attention to authority. For all that some interviewees reiterated the claim that science is a relentlessly sceptical and self-critical endeavour, they also acknowledged that a vast amount of time during that period of scientific apprenticeship is spent listening to and absorbing the knowledge from sources of scientific authority—teachers, lecturers, and textbooks.[22] However much science prided itself on its anti-authoritarian self-image, the fact was that it was only thoroughly anti-authoritarian once it reached a certain point near the scientific summit or frontier (and not even then necessarily). Up until that point, the apprenticeship of science demands a great deal of respect—qualified and questioning respect, but respect nonetheless—of authority.

This was reflected in the animated discussion there was around the importance of falsification to the scientific method(s). Falsification was cited as one of the five characteristics of science in the Overton ruling and was often mentioned in our family resemblance as a significant feature.[23] This notwithstanding, it also received a pretty savage kicking from a number of scientists, philosophers, and historians, who pointed out, with concrete examples and much enthusiasm, its inadequacies. Practically speaking, falsification did not play anything like as significant a role in their discipline as Popper and some subsequent philosophers of science imagined.[24] As one eminent cosmologist vividly put it, 'I think honestly very few working physicists feel constrained by discussion of whether Popper was right ... we

172   THE LANDSCAPES OF SCIENCE AND RELIGION

don't give a shit, to be honest' (#77). It played next to no role in established science, vast tracts of which were simply no longer a matter of falsification without that in any way affecting their scientific credentials.[25] Some scientific ideas, such as string theory, were not open to being falsified,[26] while others, such as those concerning the origins and nature of early hominids, were so underdetermined by the available evidence that they seemed to be continually falsified.[27] One cosmologist pointed out, with reference to the Michelson–Morely experiment conducted in 1887, how 'ambiguous' falsifiability could be.[28] Another, a philosopher of science, argued forcefully that Popper's approach pretty much got things exactly the wrong way round.[29] A third, a historian of science, explained how the idea and the emphasis placed on falsification owed much to the time in which it was developed, and in particular to the moral haze that was beginning to gather around the practice of science in the post-war period.

> It's a sneaky and weird way of describing what science does. It just doesn't describe what motivates people to do their research and it morally gets scientists off the hook. If you look at the period in which it was developed as a way of describing science, it was like atom bomb, *Silent Spring*, thalidomide time, and it was a way for scientists to go, "whoo, we're not trying to prove anything, we're not trying to establish anything, we're just trying to disprove things. That's all they do; they just disprove things". I think it's just a really invidious way of describing what science and scientists are all about. (#96)

In short, falsification was *arguably* important to the practice of science in *some* circumstances, but the idea that science as a whole could be characterised as a relentless attempt to falsify itself, an ever-restless enterprise of self-refutation, perpetually dissatisfied with and disrespectful to all forms of authority, including its own, was simply not true.

'Hypothesise, observe, experiment, analyse, explain, repeat' might be the foundational methodological framework for science, but it is insufficient in itself. Not only does science often require imagination and intuition, but also it demands (subjective) judgement, tacit knowledge, apprenticeship, and qualified attention to established authority.[30]

The actual practice of science
And that is when things are going well, when science is working as it should. The complexity and messiness of science is not only due to the fact

that its approach to acquiring knowledge demands judgement, imagination, instruction, etc. over and above its commitment to theorised and formalised method(s), but also because the actual practice of science often falls woefully short of the theory.

This was made abundantly clear by our interviewees, not least the practising scientists. At an individual level, we were told, scientists could be arrogant,[31] dogmatic,[32] defensive,[33] 'politically' partial,[34] personally invested to the point of actively biased,[35] and depressingly lacking in self-criticism. As one expert in human evolution remarked:

> I admit as a scientist that many scientists don't always operate as scientists. They spend a lot of time looking for data that reinforces their own positions, rather than finding ways of testing their own position and being open to alternative views. (#72[36])

They are, in a word, 'as *human* as the rest of us' and 'therefore, just as prone to error, and vanity, and egotism' (#73; emphasis added).

The problems are not restricted to the level of the individual. The practice of science could be riddled with biases and problems at a wider institutional or cultural level. This could be because of scientists' narrow and limited socio-economic backgrounds.[37] It could be because of the small-p politics rife within the scientific world.[38] It could be because of the fiercely competitive, publish or perish pressure of academic science today.[39] It could be straightforwardly financial, particularly in medicine.[40] It could even, it was claimed, be at a cultural–civilisational level.[41]

All this, of course, is contestable, and for all that pretty much every working scientist we spoke to (not to mention the sociologists and journalists of science) acknowledged both their own fallibility and that of the institutions in which they worked, by no means all would have recognised structural socio-economic problems with science, let alone cultural–civilisational ones. Quite how systemically fallible science is is open to much animated debate. Nevertheless, the point was that when out of its 'Sunday best', science could look like a rather mundane and untidy activity, a notably flawed and fallible collection of humans attempting, collectively, to reach some agreed notion of truth.

Crucially, however, there was a circuit break here. However human, fallible, self-serving, even dishonest scientists might be, the established processes of science could cope with it. What makes science science,

174 THE LANDSCAPES OF SCIENCE AND RELIGION

what elevates it from all the other all-too-human, fallible, self-serving endeavours that humans have engaged in over the centuries, not least religion, is its predisposition, hard-wired into its very being, to weed out such faults and failures. In the words of one philosopher, science 'is pretty well designed to root out frauds, bogus theories and so on' (#55). Put another way, however fallible individual scientists may be, science itself isn't, precisely because the collective processes of the scientific method are designed cumulatively to eliminate the errors and mistakes.

This might sound rather circular: science-in-theory is tarnished by the reality of science-in-practice, but science-in-practice is rescued from failure by science-in-theory. What elevates it from this circular logic is the recognition that the scientific method(s) that are designed cumulatively to 'root out' the errors and mistakes made in the practice of science *are embedded in formally established institutions*—the processes, organisations, hierarchies, and systems of authority and recognition—that are fundamental to science today. The science-in-theory that rescues science-in-practice is not simply the theory of 'experiment–observe–repeat' but the way in which those methods are incarnated in scientific institutions.

Historian of science Naomi Oreskes makes a similar argument in her book, based on the Tanner Lectures on Human Values at Princeton University, *Why Trust Science?*, referenced earlier in our chapter on Defining Science. In it she argues that the traditional answer to this question 'why trust science?'—the scientific method—is not enough. Quite apart from the holes that have been picked in the idea that there is such a thing as *the* scientific method, even if there were such a thing it would not exist in the world independent of scientists themselves. The method or methods of science are the ways in which people who call themselves scientists (and many who don't) organise their endeavours. That being so, it is not just those methods that cause us to trust science, but the people—in the plural—who implement them. In Oreskes' words, 'it is not so much that *science* corrects *itself*, but that *scientists* correct *each other*'. If this works, it works well. 'Objectivity [emerges] as a function of community practices rather than as an attitude of individual researchers.'[42] If it doesn't, science loses its trustworthiness.

Oreskes goes on to answer her own question by outlining five 'themes' that combine to produce trustworthy scientific knowledge. One is method and another is evidence, but, tellingly, the other three are consensus, values, and humility, 'themes' that direct the question of why trust science away from its narrow methodological confines and towards the wider social, ethical, and institutional context in which it takes place. In effect, science corrects itself by scientists correcting each other, and we can trust scientists to correct each other because they are not simply operating as autonomous, free, but fallible individuals, but as apprenticed, accredited, authorised, connected, disciplined, and respected agents within a wider, formalised system. The success of the method(s) of science cannot be divorced from the institutions of science in which those methods are embodied.

The scientists (and sociologists, philosophers, and journalists of science) to whom we spoke echoed this sentiment. Science's remarkably successful approach to acquiring knowledge was heavily dependent on its institutional health. Institutional here could refer to a number of things. One interviewee, with a special interest in science communication, delineated two 'layers' of institutional belonging. At the broader level, scientists felt 'a sense of identity of belonging to a wider community ... institutionalised in terms of norms and practices ... that bind [them] together no matter how disparate their actual practices' (#81). At a more specific level, scientists operated within 'actual organisations' that imposed their own particular norms and practices. These were primarily but by no means exclusively universities, although government, the private sector, and the charitable sector all also employ scientists and impose their own norms—albeit norms that are not equally trusted, if the PAS 2019 data quoted in chapter 6 are to be believed.

Across both these 'levels', there were specific institutions that helped secure the health of science. Qualifications, awards, grants, formal titles, articles, journals, editorial boards, the entire peer review process, management committees, ethics committees, the Research Excellent Framework: all of these were fundamental to the methodological success of science. Chafe and complain as some did, it was institutions of this kind that maintained the credibility of scientists, their work, and ultimately the body of knowledge that the practice of science amassed.

So, we have an understanding of science as an institution, characterised not only by a particular set of methods, but also by its own offices, structures, authorities, procedures, rewards, penalties, and codes of behaviour, as well as by its own language, symbols, and even sometimes dress. It is an understanding that sounds oddly similar to (certain disambiguated features of) religion, to elements of Geertz's definition ('a system of symbols'), or

176 THE LANDSCAPES OF SCIENCE AND RELIGION

perhaps to Smart's legal and social/institutional dimensions of religion. It certainly struck a number of our interviewees in this way. According to one sociologist:

> [Science is] like a church ... it has its own methods, it has its own rituals, it has its own hierarchies, it has its own reward system ... It's a social group that you're socialised into, and it takes a long time to be socialised into it and its premises and its methods and its approaches and its prejudices. (#67)

This was a provocative comparison, and it was not always received or deployed positively.[43] One of the more historically literate interviewees pointed out that it was this kind of institutional competition that lay behind (some) moments of science–religion tension in history, particularly in the later nineteenth century.[44] Another traced that tension up to today. 'I do think they are using science as a kind of surrogate for religion, the scientists are the new gods, and they have all of the answers and they can save us' (#93). We will return to this potential for socio-political tension in a later chapter.

In as far as there was tension here over which institution should exercise social authority, it was borne of the methodological correspondence that this wider understanding of the scientific method unearthed. It was precisely because science's method towards knowledge involved elements—imagination, judgement, apprenticeship, authority, institutionalisation, processes, etc.—that were discernible within religion's method of belief formation that there could be tension between the two.

This broadening of science's actual methodology was acknowledged by most interviewees, including the practising scientists. That said, some elements were contested and nowhere was this more so than the question of whether this full, institutionalised understanding of science's methodological success in any way involved 'faith'. Was faith an intrinsic part of science's overall methodology? A few thought it was, out of necessity. To be a practising scientist, even the greatest of them, meant having beliefs.[45] It meant placing faith in the methodological steps and institutions that you have inherited.[46] It meant trusting peers to help you navigate the institutional complexity of the system.[47] It meant trusting scientific authorities, whenever you strayed out of your particular field of expertise.[48]

More, however—indeed, a clear majority—did not accept this. For them, 'faith' was a potent trigger word. 'I think if you use the word "faith", a lot of scientists would be very angry, so I wouldn't … I think it's a leap of faith, but that's not really what … it implies a lack of evidence, you're making a guess' (#4). It was best avoided when talking about science.

> My maths isn't great, right, so I have not sat down with the mathematical descriptions of the Big Bang, or of the standard model, or of things like Higgs field. I am pretty much taking it on faith. Faith is a bad word here actually. I'll take that back. But I am definitely taking it on trust that the scientists who did the maths, and the experiments knew what they were doing. (#40)

Trust was fine. Faith was not. Science needed trust—in the institutions and processes that helped maintain the health of the scientific method and protect it from the inevitable biases and fallibilities of scientists themselves. But it did not need faith because, for many people, 'we take faith to mean a belief of things regardless of the evidence for them' (#1).

Revisiting religious methodology

Is that how religious believers conceive of faith? And, more generally, is the religious approach or method (if such a word is warranted) to acquiring and holding beliefs accurately captured by this understanding of faith? In much the same way as a careful examination of the way in which science actually operates reveals a more complex picture than theorised accounts of the scientific method often paint, so it does with religion. Just as, methodologically speaking, science is more than hypothesise–test–repeat, so religion is more than hear–believe–submit.

There is, first, a highly salient caveat to enter at this point. However subtle the religious approach to acquiring beliefs can be, it often isn't. Interviewees talked with feeling about authoritarian Catholicism, American Protestant fundamentalism, and Hindu nationalism. For all that such movements could be more variegated than they were pictured, there were, and there are, many genuine examples of extreme methodological simplicity, of people believing 'just because …'. On a personal level, interviewees told stories of how, for example, a religious person, having been backed into an intellectual corner by their interlocutor, escaped only by revealing their 'true' methodological colours. 'He realized that that was a dodgy question and so at this point he said, "Well, it's rather simple … I believe in the word of God

178 THE LANDSCAPES OF SCIENCE AND RELIGION

and the word of God is in this book"' (#20). The 'religious person'—an ill-defined figure who appeared to play the same rational game as scientifically minded thinkers, only to brandish 'faith', 'Church Teaching', or 'the word of God', as if it were some unanswerable methodological trump card, once they found themselves losing the game—was a real concern, and did an awful lot of work in some imaginations.

However accurate this picture might be, there was less agreement on how typical it was. Some were happy to sweep all religious believers away into this methodological trash can.[49] Most were more cautious, recognising that only a minority of believers were the kind of fundamentalists whose scriptures or teaching short-circuited *all* the hard work of knowledge acquisition. A few were honest enough to recognise that there was something self-serving in this image of the 'religious person', and that all too often we like to 'define our own enlightenment by how stupid somebody else is' (#38).

In a sense, though, the answer to this question—whether the simplistic methodology of 'hear–believe–submit' (or 'faith', 'authority', 'revelation', etc.) was a majority default position or a minority fundamentalist one, or something between—is beside the point. Whether all, some, or vanishingly few believers acquired and retained beliefs in this way, to compare the methodology behind the beliefs of the 'rank-and-file' believer with that of the professional scientist was another example of the descriptive–normative anomaly outlined above. Indeed, even to compare the processes underlying religious belief with what scientists actually did, in all its messy, human, methodological complexity, was to make an asymmetrical comparison.

The reason for this is that scientists are responsible for *forming* the beliefs that comprise the body of scientific knowledge, whereas religious believers are not responsible for generating the body of religious belief. They may accept and assent to it, in the same way as members of the general public accept and assent to scientific ideas and theories, but they do not generate it and are not, for the most part, responsible for its defence and development. That job falls to theologically informed thinkers, councils, teachers, authorities, and the like. If a comparison with the methodological approach of scientists is to be made, it should be made with these individuals and institutions, rather than with believers 'in the pews'.[50]

This was illustrated by one of the most commonly and forcefully repeated sentiments in the whole research, namely that religious dogma (or doctrine, theology, propositional belief, etc.) was not very important to most believers. Theologians, religious studies specialists, and sociologists of religion were all unanimous in this view. Even if you were sure (as many were) that religion comprises more than simply its ritual/ethical/practical/communal dimensions, that still doesn't mean that the factual or dogmatic elements are central.

People make 'post hoc justifications' for their beliefs[51] (and not only religious people[52]). Beliefs are not primarily founded on argument.[53] 'People's religious beliefs are ... amazingly inconsistent and incoherent' (#13). Most religions are not really propositional at all. Even Christianity, the archetypally formalised belief system, is nothing like as narrowly propositional as sometimes claimed.[54] Indeed, this propositional approach only came to the forefront in the twentieth century, and then only in various forms of Protestant Christianity, and even then only partially, as the rapid growth of much more experiential forms of charismatic Protestantism showed.[55]

In short, it made little sense to compare the methodological approach of ordinary religious believers with that of professional scientists, and the speciousness of that comparison could be seen in the apparent cross purposes around the word 'faith'. To the scientists who were so triggered by the word, 'faith' meant assenting to a theory without bothering to reason, test, or verify it. Understood this way, faith was inimical to science, albeit the concept could be smuggled back in, if necessary, under the guise of 'trust'.

To religious believers, 'faith' was not primarily understood or used that way. Rather than mere cognitive assent, it was existential, as much about what went on in your heart and your life, as in your mind.

> The notion of belief is a very western construct, in the sense that we tend to think that belief is what goes on in your head ... I am not saying it was not intellectual because it obviously was for many people, but it was about a way of being in the world. By that, I mean physically, bodily, materially, socially. (#63)[56]

In this way, the concept of 'faith' fell into the chasm that opened between normative and descriptive, the word meaning one thing on one side and another on the other. It serves as an example, and as a corrective to the idea that one can straightforwardly compare the methodological approach of science (even in its extended, institutional, and sociological fullness) with the approach adopted by religion.

And yet, even if a straightforward methodological comparison between science and religion is fallacious, that doesn't mean that there is no comparison to be made here. 'A propositional approach to the divine realm is not central [but] I wouldn't want to dismiss altogether' (#75). Unless we

180 THE LANDSCAPES OF SCIENCE AND RELIGION

are willing to embrace a wholesale social/ritual understanding of religion, we will come up against the fact that religion is based on a certain set of beliefs about reality. The Creeds, Articles, and Confessions of the Christian church(es) may have been borne of their particular contexts in late antiquity or early modern Europe, as opposed to being original and essential to the faith, but they nonetheless *were* developed and they did (and do) remain central to worship for centuries. Dogma may not be very important to most (to some? to many?) religious believers today, but that doesn't mean it's totally irrelevant. Faith may be existential, but that doesn't mean it is not at all propositional. As one philosopher told us, 'religions do have content, they have descriptive claims, and they have prescriptive claims' (#61).

In short, there *is* a reasonable methodological comparison to be made here, but it is between science and *theology* rather than between science and religion. The legitimate comparison is between the ways in which theologians and religious authorities develop, modify, justify, defend, clarify, and systemise their beliefs, and the ways in which scientists develop theirs. And that being so, the presenting question is really: how different, how dichotomous, are these two methodological approaches? Because if *theology's* methods are blunt and simplistic—if it is no more sophisticated than revelation plus tradition equals faith plus obedience—then it doesn't matter how much we finesse and broaden science's methods. There is still a fundamental tension here.

For some interviewees, as we have already noted, this was indeed the case. No matter whether you dressed it up as 'theology' as opposed to 'faith', all forms of religion were methodologically impoverished. For many more, however—for the sociologists of religion, religious studies specialists, and theologians to whom we spoke, as well as many of the philosophers and scientists—theology did have a recognisable and respectable methodological approach, even if it was one whose conclusions they repudiated.

The intellectual or theological approach to religion has its own set of methods, comparable if not identical to those of science. It has an evidential basis that includes both textual and empirical evidence. Within the textual category sit accounts of the historical events on which the religion was founded, the ritual and ethical laws and teachings that comprise the scriptures from which the religion takes its ultimate cue, and the imaginative stories intended to convey deep truths about the nature of God, humanity, and the universe.[57] Within the empirical category sits evidence derived, for example, from observing the ways in which religious beliefs were experienced and expressed in ritual, communal, and ethical practices.[58] The theological method would assess evidence from both sources, and deploy critical reasoning in its analysis, interpretation, and systematisation of the ideas and theories (one interviewee even used the word 'hypotheses' here[59]) that are drawn from them.

With regard to textual data, the 'science' (the word is not, we think, out of place) of hermeneutics was foundational. This involved serious attempts to understand the text in its fullest historical, philological, linguistic, literary, socio-political, and cultural contexts. It meant not only grasping the reality of metaphorical language (a feature that is hardly absent in science[60]) but also simultaneously refusing the simplistic dichotomy between literal and metaphorical; metaphor is rarely just *mere* metaphor. It could even involve quantification and statistical analysis, such as when attempting to understand the relationship between surah length and chronology in the Qur'an, or the true extent of the Pauline corpus in the New Testament.[61]

> I don't think it's simply a straightforward case of saying here's what the text meant in the ancient world, therefore it must mean the same things now. To me, it's kind of here's what the text meant in the ancient world and here's a history of interpretation and here's how we try to make sense of the world as [we] go along in some very different contexts. (#65)

With regard to religious experiences, critical attention could be directed to the context, nature, and consequence of the alleged experience, accompanied by an attempt to interpret it within the framework of existing doctrine. Of course, the metaphysical presuppositions underpinning such a critical assessment might be fundamentally different to that of a different scientific (e.g. psychological) interpretation of the same data. Bluntly, the theologian might allow for the possibility that the alleged experience is authentic— in effect, that it is a genuine 'spiritual experience'—whilst the psychologist might deny that possibility a priori. The point, though, is not that the two critiques would arrive at the same conclusion but that both could have their own methodological integrity.

This kind of serious methodological approach to the evidence provided by religious texts and experiences is not exactly the same as that outlined in different accounts of 'the' scientific method, but it is recognisably similar. Moreover, such an approach could result, contrary to popular view, in a conception of religion that was close to science in as far as it was (1) developmental and (2) critical towards authority.

182 THE LANDSCAPES OF SCIENCE AND RELIGION

Taking the first of these: like it or not, religious beliefs did change and develop. In the words of one theologian, 'what theology is doing, and has been doing these 2,000 years, is recalibrating its beliefs according to ... our current state of knowledge about the world' (#88). The extent, direction, and validity of that change is hotly contested, but both sympathisers and critics acknowledged the reality of such religious 'progress'; they simply differed in their interpretation of it. For the former, it was a sign of intellectual vigour or even divine guidance. 'We've seen how religions have evolved. St Newman argues that there's an evolutionary development of religion under the grounds of the Holy Spirit' (#41). For the latter, it was a sign of painful hypocrisy, the irreconcilable claim to know The Truth while simultaneously changing it. 'Many religions like to paint themselves as never having changed position, whereas even a cursory glance at the history of many, you'll see a very clear movement and the internal battles within them to adapt to circumstances and the new political realities' (#62).

With regard to authority, the reality of a genuine theological attention to and critique of foundational texts and religious experiences meant that, at the extreme, religion could be inherently antagonistic to both written and institutional authority.[62] Less extremely, it meant that there was liable to be more genuine space and flexibility when it came to religious authority, textual and institutional, than usually imagined. A number of interviewees independently referenced their surprise when they had actually met or dealt with Catholic authorities and theologians, discovering (to their astonishment) that there was considerably more intellectual freedom than they had been led to believe. In the words of one science journalist, 'I remember listening to theologians during the [Liam] Donaldson report [into the cloning of embryos] and I was always amazed [that] even among Catholic theologians, you can have ten very different views, but they're all Catholic theologians' (#50). Several others echoed this view.[63]

When considered in this way, the methodology behind theology was not just blind faith, an automatic adherence to authority, or an unreflective response to self-designated spiritual experience. To be clear, and at the risk of repetition, religious faith at a popular level *could* be all of those things. But that didn't invalidate it any more than popular belief in the Big Bang, quantum mechanics, or the theory of natural selection was invalidated because people unquestioningly accepted them without being able fully to understand, let alone mathematically defend, them. For the methodological comparison to be a fair one, it has to be between two comparable activities.

If we understand 'the' method that underpins scientific activity in its fullest and truest sense, we will understand that it requires more than simply a mechanical application of certain established steps. Science requires imagination, experience, and judgement; respect for authority and a commitment to certain ethical values; and recognition and adherence to titles, institutions, and processes. In a similar way, the process by means of which religious authorities and scholars acquire, refine, deepen, defend, and develop religious beliefs shows a methodological approach that pays serious critical attention to relevant textual and empirical evidence, and is prepared to develop, revise, and abandon 'theories', as well as question, challenge, and change established and authoritative views on the topic.

Understood this way, the methodological tension between science and religion is diminished. The complex and variegated approach of both science and theology to acquiring and holding beliefs appears to be not as categorically different as all that. But if the gap is diminished, it is not closed—careful attention to the methodological approach of both can draw out their similarities, yet some tensions remain.

Outstanding questions

For all that a close examination of respective scientific and theological approaches to acquiring, holding, and refining beliefs may reveal parallels and resonances—about evidence and experience, critical analysis and subjective judgement, induction to the best explanation and institutional authorisation, even quantification and respect for authority—it is clear that the two approaches are not identical.

Experimentation, falsification, and prediction, important if not essential to the methods of science, are largely absent in those of theology. Conversely, attention to human imagination (such as is elemental to written texts, even those revered as divine in origin) and respect for the principle of revelation, both of which are part of the theological enterprise, are (entirely?) absent within the practices of science. As noted above, we can diminish the methodological conflict, but not close it altogether.

Part of the reason for this is natural and, as it were, wholly innocent. Science and theology study different things and, to adapt a phrase of which the theologian–cosmologist John Polkinghorne was fond, ontology informs methodology.[64] We naturally investigate and analyse different things in different ways. The principle applies to different disciplines within science[65]

184 THE LANDSCAPES OF SCIENCE AND RELIGION

as we saw in part I—'if you want to say that particle physics and psychology sit in the same realm of certainty of knowledge you're completely out to lunch'—so it is only natural for it to apply to a still wider range of intellectual disciplines.

Science's success, as we have noted, is borne not only from its methods but also from cognisance of its limits. As the *Dover Area School District* submissions pointed out, science does not consider issues of meaning and purpose. For a minority, this is because meaning and purpose are illusory. For others, it is simply because such things are not amenable to the normal methods of science but are to be studied by other intellectual disciplines, such as philosophy or the humanities. Just because those disciplines do not conform to (all) the methodological features of science does not make them less intellectually respectable.[66]

This point was made repeatedly by interviewees, including many practising scientists. Qualia,[67] values and concepts such as love or justice or ethics,[68] unique human events,[69] products of the imagination,[70] history (or at least particular historical events)[71]: science struggles with all of these. To be clear, and to return to the principle of a family resemblance understanding of science, it is not that the study of any of these areas is necessarily *un*scientific, or that it fails to share *any* feature in common with science. Most will involve some form of observation, evidence, and critical analysis; and some might even integrate a measure of quantification or falsification. That might qualify them as honorary sciences, distant but recognisable cousins of physics and biology. We must recall Wittgenstein's rhetorical question, 'What still counts as a game and what no longer does? Can you give the boundary?' and his firm 'No' in response. If there is no firm boundary condition, we cannot definitively say that *x* discipline is not a science. We can only say that its resemblance to practices that we *do* comfortably call science is faint to the point of invisible.

The key point in all this is that the disciplines that do study qualia (or justice or ethics or unique historical events or products of the imagination) *are* legitimate intellectual enterprises despite only sharing perhaps a few of science's trademark methods—and there is no reason why theology should not also fall into this category. As several interviewees pointed out, political theory or constitutional studies, for example, wrestle with many of the same issues as theology does: textual uncertainties, ambiguity of language, opaqueness of original intention, the need to reflect changing circumstances, etc. They can even replay quasi-religious battle lines, such as between originalism and living constitutionalism, or different schools of Marxist or liberal theory.[72] None of this means they are in any way betraying their methodological integrity. Theology, methodologically speaking at least, falls within this category.

All this means we should *expect* some methodological difference between the practices of science and theology. 'Religion is something that is essentially tied up with what it means to be human and questions of meaning and morality and redemption and love, all of these categories are not scientific categories and to try and squeeze religion into that straight jacket is really distorting' (#88). Nevertheless, there are still some questions about the remaining methodological space between science and theology that merit attention. We would like to mention four—pertaining to revelation, evidence, consensus, and confidence—as a way of concluding this chapter.

First, revelation. Does the methodological approach of theology ultimately exclude from scrutiny some sources of evidence, whether textual or empirical, because they are deemed, on principle, to be foundationally authoritative, i.e. revelation? It is, of course, true that the foundational texts of a religion, be they historical, literary, or legal, are, in theory, open to intense scrutiny. Hermeneutics is a serious discipline, as is the interrogation of alleged religious experiences. However, there is at least the potential for tension here, if that theology also takes that evidence to be foundational or finally authoritative; if, in effect, its approaches a text or an experience as a priori a revelation.

We need to be clear here. The problem is not having certain basic beliefs or presuppositions that are judged as authoritative without having, in a phrase we have quoted before, 'wonderful epistemic warrant'. Science has such commitments, as we pointed out in part I. Indeed, it needs them. The challenge comes from, on the one hand, having such foundational beliefs, while on the other claiming that they are simultaneously open to interrogation. Science has basic commitments, such as the existence of laws of nature, the intelligibility of the universe, and the reliability of human senses. It assumes them, needs them, builds on them, but it doesn't claim to be able to interrogate them. Theology has others, such as the belief that certain texts or experiences are fundamentally revelatory. But if that commitment to revelation underpins the theological approach in the first place, in much the same way that belief in the laws of nature underpins a scientific approach, can we also claim that those texts and experiences are fully open to scrutiny? Does willingness to recognise the category of revelation take theology outside the city walls of the academy? In the words of theologian William Abrahams, 'if truth be told, the contemporary academy does not find the appeal to divine revelation at

186   THE LANDSCAPES OF SCIENCE AND RELIGION

all attractive. Outside theology, and often within theology itself, the appeal to revelation is simply not permissible'.[73]

In effect, the question lies with the concept of revelation—not whether revelation is possible in itself (that question lies beyond our purview here), but whether granting something (e.g. a text or an experience) the status of revelation ultimately puts it beyond scrutiny, and means that the methodology inherent in theology will always fundamentally be different to that of science. In the words of one philosopher, 'I have what I regard as rational grounds for accepting the authority of science because I have some understanding of what scientific method is ... it's quite different from the idea of a religious revelation' (#5). In short, might the concept of revelation be a fundamental sticking point between science and religion?

Second, evidence. The evidence treated by theology is often derived from personal experience, either immediately or behind the texts that bear witness to the religion's origins.[74] For some, this is anathema to 'real' science,[75] and although this position might be deemed an unduly narrow restriction— plenty of intellectual disciplines, some scientific among them (e.g. anthropology), take 'subjective experience' seriously—there is a problem with taking experience as legitimate evidence.

The problem is not that subjective experiences can be illusory, open to suggestion[76] and misinterpretation,[77] or even hallucinatory or hysterical.[78] It is quite possible to believe experience to constitute, in principle, legitimate evidence, and yet simultaneously to judge, after interrogation of specific experiences, that the claims made on the back of them are false. The real, *methodological* issue here is how such judgements can be arrived at. Readers will recall that one of the features of science identified in the process of disambiguation described in part I pertained to its subject of study. It was the vaguest of the six outlined. People were clear that science could be characterised by its subject; they were just unclear how that subject should be described: nature, the material world, etc.

Among those descriptions was the idea that the subject should be 'in principle observable', something 'to which humans have common access'. For the method(s) of science to get going, what they work on has to be accessible to more than one observer. If not, there is no comparison to be made and that means there is no opportunity to develop any inter-subjective or quasi-objective view.

In light of that, the weight placed on subjective experience as evidence can be problematic. It may well be, as one sympathetic philosopher put it, that 'in the case of religion, evidence is never of the kind that can easily be infused into another mind if that mind is not already aware of what it is you're talking about'. But, if so, it follows, as the same philosopher acknowledged, that 'argument is almost fruitless because by definition whatever one is talking about is greater than anything that you can formulate propositions about' (#75).

Once again, we should stress that this is a narrowly methodological point. The issue is not whether spiritual experiences are real, or whether subjective experiences can be admitted as evidence. The methodological point of tension arises around the specific question of how, in the absence of common access, they can be satisfactorily be evaluated. In the words of one sceptical chemist, if 'they believe in God ... because of their own personal experience, then they're going by evidence which they have, which I don't have. And they're coming to conclusions which they accept but I'm not having' (#46).

This point about evidence leads on directly to a third, about consensus, or, more specifically, the processes for disciplining the conclusions of theological enquiry. The fuller and more descriptive understanding of how the scientific method(s) functions pays full attention to the way in which authority and institutions play a seminal role. In the light of that, the role of authorities and institutions in the development and disciplining of theological beliefs doesn't seem so alien or threatening. Those who like to point their polemical finger at the Vatican for its proscription of theologians easily forget those scientists, like Woo-Suk Hwang, Diederik Stapel, Marc Hausser, and Jan Hendrik Schon, who have been disciplined, struck off, or in some cases imprisoned for bad practices.[79] Ultimately, both science and theology have *and need* an institutional infrastructure to legitimise practices, winnow findings, discipline practitioners, and the like.

And yet, in spite of this similarity, there remains a problem. However much scientific institutions are called in to discipline scientists, the subject of study does so first. Scientific disciplines have a habit, albeit over time and with a great deal of noise and disagreement in the process, of converging on agreed ideas. They may (in theory) remain tentative and open to falsification, but in practice, as we heard, they are accepted as true. Reality disciplines scientists before their institutions do.

It is not that this never happens in theology. Agreement about a source of evidence (e.g. a particular text) and the process of careful analysis (e.g. hermeneutical principles) can equally result in convergence on certain ideas. However, this history of theology shows how this is by no means necessarily the case, and how the eventually agreed solutions have often required

188   THE LANDSCAPES OF SCIENCE AND RELIGION

the disciplining interventions of theological institutions, which has usually meant ecclesiastical, and sometimes political, authorities.

Once again, the issue here is methodological. How far do theologians naturally converge on agreed beliefs? How is consensus arrived at, and on what basis? Who decides what is true, and how? Such questions should not be taken to imply that there are none. Internal incoherence would be one. Failure to reflect and respect properly the accepted sources of evidence would be another. Plenty of theological ideas have fallen out of favour over the years for reasons other than institutional judgement. But the challenge still stands. For all the falsification isn't essential to science, it still plays a role. Where are the sources of decisive falsification in theology?

The fourth and final point brings us back to the greatest trigger word in this whole methodological debate: faith. As we noted above, there is much feeling and not a little misunderstanding around this word, which is often used in different ways by different participants in the conversation.

That notwithstanding, it is possible to arrive at a conception of faith that doesn't simply exacerbate the potential methodological tension between science and theology. The actual practice of science, as we were told, *does* rely on trust (to use the preferred word). Scientists do operate on hunches that they can't prove. Theories are often underdetermined by the available evidence. In this regard, 'faith' (the scare quotes are sadly necessary) is not alien to science.

At the same time, religious faith can be understood in an existential rather than narrowly cognitive way, as an acknowledgement that no path taken in life, religious or not, can be known with complete certainly and confidence to be true and right. We necessarily step out in faith by dint of our humanity. The paths we take are experiments in living. Approached this way, the methodological conflict threatened by certain understandings of 'faith' is closed up. Both, in their own way, follow the advice of the old Russian proverb 'Trust, but verify'.

However, it is only fair to acknowledge that 'faith' is not always used that way by religious believers. Indeed, quite to the contrary, faith is sometimes used to mean certainty beyond that which the examination of mere evidence can supply. To take one example, from a pretty unimpeachable authority on the matter, in his second popular book on Jesus of Nazareth, Pope Benedict XVI wrote, 'it is faith that gives us the ultimate certainty upon which we base our whole lives'.[80]

This is a conception of religious faith that does appear to be at odds with the more emollient, epistemically humble understanding of faith outlined above, the one that tries to close the methodological gap between science and religion. If this is indeed what theologians mean by faith, it leaves us with an awkward question. Is the methodological approach adopted by religion—or, more properly, by theology—as open and receptive as that adopted by science, at its best, is? How serious, in fact, is the methodological conflict here?

In summary, when people disagree about science and religion, one of the things they are really disagreeing about is methodology. How does science and how does religion establish and hold the beliefs they deem to be true, and what are the processes each has for defending or modifying them?

By some reckonings, there is an inevitable methodological opposition here. Science questions authority, religion relies on it. Science overturns tradition, religious builds on it. Science values epistemic humility, religion certainty. Science is organised doubt, religion institutionalised faith. And so on.

This chapter has sought to show that this difference is not as severe as all that, and that when we revisit and examine carefully both the complex methods and actual practice of science, and those whereby religious beliefs are developed and refined, we encounter a range of common phenomena: not only evidence, critical analysis, openness, inference, repetition, etc. but also imagination, judgement, apprenticeship, authority, institutional involvement, etc. The methodological conflict may not be as extensive as all that. When people disagree about science and religion, they may not need to disagree as violently as they do.

All that recognised, however, a methodological 'skirmish'—or perhaps 'discussion': choose your metaphor to taste—does remain. The approach of science and of theology to establishing, holding, and modifying beliefs is not identical. There is no reason why it should be. Different disciplines naturally adopt different methodologies. But there remain areas—we have highlighted revelation, evidence, consensus, and confidence—where genuine and relevant differences remain. There is still a conversation to be had when it comes to method in science and religion.

Notes

1. 'I wouldn't be asking anybody questions about the relationship between science and religion head on. Because apart from epistemologists, very few people spend much time thinking about this.' (#21) This is, of course, in line with John Evans' argument in *Morals Not Knowledge*.

190 THE LANDSCAPES OF SCIENCE AND RELIGION

2. 'Belief to me is the acceptance of something to be true without the evidence because if there's evidence there, you don't need belief.' (#18)
3. 'I think the fetishization of faith is a marvellous trick, to create doubting Thomas, and to say that this man is bad because he believes it to be unlikely that someone could have risen from the dead and appeared as a ghost.' (#39)
4. 'A sort of paradigmatic example of what we mean by faith ... [is] Tertullian['s], "I believe because it's absurd", the idea that is meritorious in some way to accept something even in the face of countervailing evidence, even though the evidence is against it.' (#61) The philosopher who cited him was apparently unaware that Patristic scholars had long ago shown that Tertullian never wrote those words or meant what his (Enlightenment) interpreters claimed. On the (deliberate) misquotation and misinterpretation of the phrase see Harrison, Peter '"I believe because it is absurd": the enlightenment invention of Tertullian's *Credo*', *Church History* (2017), 86(2), 339–364.
5. 'I think it's not unfair to describe parts of religious faith as deliberately and intentionally belief without evidence.' (#1)
6. 'Religion to me seemed based on faith, that one didn't, or shouldn't require proof to believe, it was a feeling.' (#8)
7. 'Religious experiences, you can explain them, things like mass hysteria ... people are very suggestible.' (#9)
8. 'Human beings throughout history, and this is not just [about] God, have created narratives and myths that suit their desires and their anxieties.' (#86)
9. 'Once you have got this notion of—not just authoritative—but a form of divine writing, scripture, that is not to be challenged, then the cerebral aspect, and the intellectual notion of religion formally—so formally in terms of religious authorities, so, clerics and priests— formally eclipses behavioural, bodily, social practices.' (#63)
10. 'Religions cherry-pick their own traditions and their scriptures, their documents in order to choose the bits that they want to continue to believe in.' (#61)
11. 'They'll cherry-pick the one bit that seems to in hindsight back up the statement they're trying to state but then ignore several other statements that say the complete opposite.' (#18)
12. For example, 'the authority is "you have to believe what I say" ... it's not that bad in the Church of England, but it can be. It is a lot worse in many of the evangelical faiths in the States and in Islam it is terrible. That sort of authority I think is crushing to the most wonderful human spirit of curiosity and openness.' (#24)
13. 'What we think of as reason is broader than a lot of more positive scientific people might say and I think that people ... sometimes ... do have reasons to their belief. I just think that, ultimately, they're just not very good ones.' (#3)
14. 'The best thing that scientists love doing is proving their colleagues wrong.' (#18)
15. 'If you go to any scientific conference, and someone will stand up and present their data, and someone else will say that is all complete shite. It is hugely confrontational.' (#29)
16. 'Religious people assume authority because they're wearing a certain costume.' (#13)
17. 'I think organised religion makes claims based on tradition and on historical record and on legacy which I constitutionally find it hard to accept purely because they're tradition.' (#7)
18. 'I rather like the motto of the Royal Society, which roughly translates to "Don't take anybody's word for it". I mean, that is what, in my view, real knowledge is about ... The difference between that and the religious view is that there will be some people who think that science is just about reading textbooks, and you know, soaking up knowledge that way, without every questioning what is in those textbooks. That is not science.' (#26) In a similar vein, 'some people try and live their lives with a book. Whereas at least science it replaces itself with new knowledge. And because it does this it allows people to have a more flexible and open-ended relationship with it' (#71).
19. 'I think that is the whole point about science, is that it shouldn't overclaim because we have this whole history of models that were thought to be correct.' (#66)

METHODOLOGY 191

20. 'What we do is talk about the evidential base. No claim is ever seen as true, and no claim is ever seen as false.' (#25)
21. See Spencer, Nick, *'Beauty Is Truth': What's Beauty Got to Do with Science?* (London, 2022).
22. This important feature of science could be defined out of existence—doing science at school and university, or at any level before you start doing original research, is not *really* science—but this seemed far-fetched, and an example of post-hoc redefining.
23. Curiously both the Science Council's definition and *Dover Area School District* (2005) plumped for the rather more demanding goal of verification.
24. 'People often bang on about Popper and falsification as being the be all and end all of science. But if you actually look at the history of how science has panned out ... often we haven't necessarily followed falsification strictly.' (40)
25. 'I don't think the atomic theory of matter is speculative anymore.' (#55) 'We don't think the law of gravity is waiting to be falsified, we know not to step out of windows, you know.' (#1)
26. Which meant, at least in the eyes of some, they ceased to be science. 'If you can't falsify it, you can't do science on it. It is a bit like string theory. I call string theorists basically religious. They are, because they can't falsify it *right now*.' (#27; emphases added) The temporal caveat here is significant.
27. 'It depends on theories about the origin of humans from early hominids, people keep on changing their minds because we don't have very much evidence. A few bits of bone and teeth lying around. There the evidence isn't strong enough to decide between the theories.' (#55)
28. 'Let me say also why the idea of falsifiability is in itself a bit ambiguous. You probably know about the Michelson–Morely experiment, this is important for Einstein, it goes the speed of light is the same however you're moving which is very surprising. This experiment was done in the 1880s and it was one of the things which were in Maxwell's theory. Suppose the Michelson–Morely experiment had been done in the sixteenth century, we'd have used the arguments against Copernicus. They would say the Earth doesn't move and so the point is you can't have a clear-cut reputation.' (#33)
29. 'I think Popper is a much-overrated answer of science. He's got things completely the wrong way around and indeed it's his view that science never gets into truth. That every scientific theory will in time turn out to be false and I think that just makes nonsense of what science is. If you ask some normal people what's the difference between astrology and astronomy or atomic chemistry, they would say that astronomy and atomic chemistry have been established by the facts and astrology certainly hasn't ... and they'd be dead right. Popper somehow argues himself into the situation that, as far as being established by the facts, they're all in the same bad boat. That astronomy and atomic chemistry are just as bad as astrology, they're just unsupported speculations. And now he has to find some way of ... recovering the difference between the sciences, the disciplines that get a tick and those that don't, and he comes up with falsification. It's all a long way round driven by his mistake that all science is speculation.' (#55)
30. 'Of course, as scientists working within a particular theoretical paradigm, we do tend, at least for the purposes of our day-to-day work, regard the hardcore propositions of that research programme, to use Lakatos's term, as basically irrefutable.' (#21)
31. 'Scientists are generally quite arrogant people, and they think they have all the answers to things.' (#13)
32. 'One of the problems we have is that scientists can be as dogmatic as any other group. Rationalists can be as dogmatic as any other group and sometimes I worry that that kind of dogmatic attitude is not compatible with good public debate and public discourse.' (#48)
33. 'Science is this group of people as a society of people, they, just like organised religion, when they have to, they will huddle together and defend themselves from external forces.' (#29)

192   THE LANDSCAPES OF SCIENCE AND RELIGION

34. 'I actually do think there is a lot of politics even in science, where people aren't really going on the basis of what they're actually finding, but what they hope to find.' (#71)
35. 'A lot of that is they are bringing their own biases to the table. They are bringing their own interests and passions to the table.' (#53)
36. The connection here with anti-Popperian approaches in science will be obvious: 'Karl Popper's falsification saying that science proceeds by falsification. I mean that can only be said by someone who has never published their own theory in science because the last thing that a scientist wants to do is publish a theory and then immediately start to look for falsifying evidence from what you have just published.' (#16)
37. 'Scientists ... tend to come from a very small subsection of society, a privileged one and the white males are represented, you'll have fewer perspectives from non-privileged, socio-economic, racial, gender groups.' (#78)
38. 'The politics of getting a grant. The politics of having your work reviewed. The politics of who is the student that you kind of really give a pat on the head to and bring through into the next generation of researchers.' (#97)
39. 'So the communal consensus basis of science in our modern institutionalised version, we have all these pressures for competition that lead to cheating and fraud and all of that.' (#81)
40. 'Who is funding the science? With medicine, it's practically all funded by pharmaceutical companies, so it's not surprising that we end up trying to treat everything with drugs.' (#36)
41. 'I would argue that our understanding—western empiricism—is just one form of understanding of the world that can act as an exclusionary tool when it comes to other forms of knowledge that have been considered important in other societies. And ultimately degrades those perceptions, wisdoms, truths that could be derived through other means of understanding the world—by holding them to a standard which is not their own.' (#59) This interviewee went on to reference the work of the Nigerian–Finnish journalist Minna Salami as an examples of 'reclaiming alternative forms of understanding of the world' from the 'the euro patriarchal framework'.
42. Oreskes, *Why Trust Science?*, p. 52.
43. 'Science is this group of people as a society of people, they, just like organised religion, when they have to, they will huddle together and defend themselves from external forces.' (#29)
44. For more details on this historical period see Spencer, *Magisteria*, part 3.
45. 'Scientists are driven by beliefs as well. You look at Einstein and special relativity, he was obviously driven by philosophical consideration, he wasn't influenced by the Michelson Morley experiment as people like to say.' (#50)
46. 'There's also a degree of faith and belief involved in science because whatever system of science that you're practising requires you to trust that that method will work, and agree with its assumptions.' (#48)
47. 'If I'm looking to read scientific journals and I want to find a good scientific journal, there is absolutely no way for me to distinguish, say there are 50 different journals on the same topic. I have found that frustrating that there's no kind of benchmark. I've asked scientists, very eminent scientists to help me and how do you do it? Is it the number of citations? But it isn't, it's just word of mouth really and people know which are the good and which are the bad journals.' (#87)
48. 'I said that science isn't faith-based, but it is. Because I haven't read a paper on global warming. I haven't looked at the studies. But I believe in global warming because everyone tells me to believe in global warming. That is the authority in my current world, in the same way that 200 years ago, the authority would be the church.' (#38)
49. 'Religious people definitely don't understand that that's what science is ... what would a religious person say about that, I couldn't care less. Because I don't think they have anything useful to bring to the debate.' (#13)
50. We are conscious of the distinctly Christian language and framework at play here, which is primarily due to the way in which Christianity was the implicit model for religion in

METHODOLOGY 193

the minds of many interviewees. That said, what they did say on this matter would apply, *mutatis mutandis*, to many other religions.

51. 'We know that humans very frequently formulate *post* ad hoc explanations for things they've already committed themselves to for one reason or another, and you see this in every area of life like politics and doctrinal commitments and ... really every area of life.' (#22)

52. 'I can imagine non-religious people might invoke cosmology or Darwinism as a justification for not being religious. But I would bet a huge sum of money, if you took any of those people and investigated their biographies carefully, you would discover that that's got nothing to do with it.' (#21)

53. 'Most people who believe, if you were to ask them, would say that their belief is not entirely founded on argument.' (#42)

54. 'This idea that what's going on with Christianity say is that you sign up to a bunch of propositions and then if asked you can justify them, that's just absurd.' (#64) 'The number of religions that have propositional statements of that kind is tiny, it's tiny. [Even] Christian history was never like that.' (#67)

55. 'I guess, in Protestantism, there has maybe been a particular focus on propositional beliefs, but even in modern-day Protestantism ... I think if you study a Protestant church in actual fact, it's just as much or more about other dimensions.' (#74)

56. Or, the view expressed from the inside, as it were: 'Faith for me isn't so much the belief that Jesus rose from the dead. It is actually the way I live my life. And it is something I believe in on the basis of evidence, not just historical evidence, but I believe I have had an experience, which is a personal piece of evidence that Jesus is risen from the dead' (#44).

57. 'It's quite like science in that it's studying something, it's studying an event, a historical event and the witness to that event and the way that's been interpreted.' (#84)

58. 'People are, I think, very empirical in the way they approach spirituality and religion, they test it out, they try it out and they see if it's true ... Now it might be that you have just one or two experiences in your life that are so significant to you that you found that that's enough to base your religion on. That's not blind faith, it's just that they're very powerful experiences. They're a different sort of experience from looking at something down a microscope.' (#67)

59. 'Theology too can go out and test some of its hypothesis in the world, not all of them, but some of them it can.' (#83)

60. 'Everything we do in science is a story, right, so science, all of science is just a model of nature, right. So, atoms are not tiny little balls that are whizzing around, they're not and we know that now. But the whole of science is using metaphor and analogy to help understand and explain the physical world.' (#15)

61. On the former, see, for example, Sinai, Nicolai, *The Qur'an: A Historical–Critical Introduction* (Edinburgh, 2017). On the latter, see Mealand, D.L. 'The extent of the Pauline Corpus: a multivariate approach', *Journal for the Study of the New Testament* (1996), 18(59), 61–92.

62. 'I don't think that all religious worldviews are necessarily connected with authority. We know that many of them are ... but I think it is possible also to have ones that are based on the authority of personal experience and that's a very different kind of authority that is more to do with individualistic spiritual sense of connection to something.' (#21)

63. Another science writer said, 'One thing I have been interested ... over several years is the Vatican's response to various different scientific stuff in the news ... I quite enjoy engaging with them because they have quite a considered response to a lot of big scientific news' (#49). A third reported how, 'I went to a retreat in Guernsey with the Catholic bishop of England, and it was great ... I was just struck by their flexibility and their thoughtfulness' (#54).

64. Polkinghorne's actual phrase was epistemology models ontology. Apparently, he said it so many times that his wife bought him a T-shirt with the phrase on it.

65. Meaning, of course, those disciplines that are commonly judged to sit comfortably under the overall rubric of 'science'.

194 THE LANDSCAPES OF SCIENCE AND RELIGION

66. They will, or should, conform to some, such as careful use of evidence, critical analysis, etc.
67. 'I don't think that it can capture everything. I don't think that it can capture the way I feel when I see a sunrise or my daughter, that kind of experiential aspect to awareness.' (#36)
68. 'In that sense, the material world is all there is, but love is not something that science can investigate. Science can investigate the neuro substates of love and can tell us all sorts of things about that, so it's not that it has nothing to say about it, but love is not a scientific concept, neither is justice, so really, everything around ethics, broadly construed ethics, human contact, human value.' (#3)
69. 'You can't bring somebody into the laboratory and ask them to get divorced again, so that you can have your sociological study of divorce. Those phenomena are one-off. They are unique and so the method that you use to try to understand them is not going to be quite the same as the endless repetition of experiments.' (#61)
70. 'I would draw a distinction between disciplines that are based on that sort of evidential methodology and disciplines that are more based on … imagination as the source of knowledge … I would say the humanities are obviously primary because we're human beings and the humanities are what use our particular cognitive apparatus.' (#67)
71. 'I think of history as being an evidence-based subject, but it's not scientific. And one of the areas that has been problematic when geneticists have effectively become historians is a rather understandably arrogant view that we now have evidence which is molecular biological and very quantifiable, and that supersedes other ways of knowing the past. And I do not think that is correct and I think we've made a lot of mistakes, my community has made a lot of mistakes, in not recognizing that history has a way of knowing using its tools which have been around for significantly longer.' (#35)
72. 'It is not just religion. We are finding this with the US constitution, aren't we? The US constitution is this kind of sacred text that was written 200 years ago, and it is completely incompatible with the way that we live now. The right to bear arms, for example. If the founding fathers had known that people would invent machine guns, they never would have allowed them to bear arms.' (#29)
73. Abrahams, William, 'The offense of divine revelation', *Harvard Theological Review* (2002), 95(3), 254.
74. 'Religion ultimately relies on experience. Jesus didn't become enlightened through studying texts … Buddha didn't become enlightened through doing a PhD. Both of them achieved their authority from their own direct experience … There are large numbers of people in the world today, even some surveys show the majority of the population who had mystical experiences of one [kind or] another.' (#41)
75. For example, 'it needs to be verified and not just a subjective experience' (#1).
76. 'Religious experience can be induced by certain conditions in the brain or the lesions and hallucinations look like religious experience, but we now know these types of experiences are psychopathological.' (#101)
77. 'There is really good, solid literature of people having near death experiences where they have this kind of vision, or they have dreams, or they have memories, or whatever it is. Unfortunately, one in a hundred of those talks about an angel. And those are the ones that get picked up and discussed. You know, neuroscience has proven religion because someone saw an angel. You know, if someone saw their grandmother who has been dead for 15 years, that wouldn't be an issue. And because religion is so important to some people, it is entirely natural that in those states, that if they had a near death experience, they would see something like that.' (#15)
78. 'Religious experiences … you can explain them, things like mass hysteria in something, people are very suggestible and there are lots of incredibly weird things that the human body can do.' (#9)
79. See Chevassus-au-Louis, Nicolas, *Fraud in the Lab: The High Stakes of Scientific Research* (Cambridge, MA, 2019), for more on these cases.
80. Pope Benedict XVI, *Jesus of Nazareth: Part Two, from the Entrance into Jerusalem to the Resurrection* (London, 2011), p. 105.

9
Anthropology

Introduction

The contents of science—its actual theories, discoveries, achievements, etc.—do not feature much in the various formal and disambiguated definitions we explored in part I. Legal and institutional classifications say nothing about science being defined by what it has established about, for example, the origins of the universe, the structure of matter, the processes of chemical bonding, or the development of life. Science does 'accumulate substantial observational or experimental [data which] support scientific "theories"' (*Edwards v. Aguillard*), but the content of those observations and theories is not part of the definition of what science is. They are scientific on account of the way they are developed and held, not on account of what they say.

It is the same with the disambiguated family definition derived from our interviews. According to these, science is a series of methods, involving the study of natural (etc.) phenomena, predicated on certain metaphysical presuppositions, and characterised by particular objectives, by a commitment to certain values, and by a formalised, institutional structure. It is not the fact that the universe is 13.7 billion years old, that life developed through a process of natural selection, or that there are (currently) 118 identified chemical elements.

And yet, in some measure, that is precisely how popular opinion does conceive science. As we noted in chapter 6, the general UK public associates science most often with the biology, chemistry, and physics they encountered at school, and then with progressive narratives, many of which have a specific content to them. Science is 'drugs', it is 'cures for diseases', it is 'medicine', 'rockets', 'astronomy', and 'technology'.

This sets up another area for potential science and religion tension because religion *is* understood, both in formal and popular definitions, to be about something. Religion has content. Except perhaps for those who favour a solely ritual conceptualisation of it, religion is normally understood to hold certain things to be true. Most obviously, or at least most frequently, this means the existence of God, or some comparable supreme being, higher

196 THE LANDSCAPES OF SCIENCE AND RELIGION

power, or transcendent dimension, but it also includes a rich tapestry of ideas about the soul, the afterlife, and some form of post-mortem existence.

This could be judged a purely metaphysical issue. After all, few people think that 'there is a God' is the kind of statement amenable to empirical verification, or even falsification. If there is a disagreement here, surely it is over the presuppositions that different parties bring to the discussion, and should therefore be located in the metaphysical region of the science and religion landscapes.

This is true but only up to a point. What emerged from both the elite and general public interviews we conducted was that, however much different parties might be arguing about underlying metaphysical questions of super/naturalism, materialism, etc., they were also disagreeing (or there was, at least, the potential to disagree) about more empirical claims. Science, at least in the popular conception, had a certain picture of the way the world was, and religions had strands of dogma that also claimed that the world was a certain way.

Here, then, is a third potential 'battlefield' stop in our landscape tour of science and religion. Having paused at metaphysics and at methodology, we are now in the region of empirical content. People find disagreement between science and religion not simply on the basis of how they come to hold their beliefs, or what presuppositions those beliefs are based on, but on the content of those beliefs itself.

Content, however, is a rather vague term. Allowing that some people do find themselves in disagreement over the respective content of science and religion, the question is, what content? What exactly are people disagreeing about here? We can answer this question, albeit superficially at first, through the quantitative study that was part of the Theos/Faraday/YouGov research. This asked respondents which scientific disciplines (and, in one instance, which theory) they thought made it hard to be religious. Some were clearly more of a problem than others. Twenty-four per cent of people agreed that 'climate science' made it hard to be religious, compared with 44% who disagreed. The balance was roughly similar for psychology (25% agreed v. 40% disagreed), chemistry (28% agreed v. 37% disagreed), geology (28% agreed v. 35% disagreed), neuroscience (28% agreed v. 39% disagreed), and medical science (32% agreed v. 40% disagreed). In general, only a minority, albeit a sizeable one, saw these disciplines as posing a threat to religion.

Two other disciplines were more of a challenge. First was cosmology. When it came to the disciplines of astronomy and cosmology, 36% of people thought they made it hard to be religious, with the same percentage disagreeing. The perceived severity of this threat was confirmed when the same question was asked about the theory of the Big Bang, which was deemed to make it hard to be religious by 39% of people, with only 31% disagreeing.

Second was biology, or more specifically evolution. Research conducted by Fern Elsdon Baker, Amy Unsworth, and others has shown that anti-evolutionism in the UK is complex, subtle, and not as bluntly oppositional as it is in the US.[1] Moreover, according to Elaine Howard Ecklund, even in the US it is rather subtler than it is often given credit for. That recognised, even if explicit religious rejection of evolution on, say, scriptural literalist grounds is relatively rare in the UK—according to Theos/Faraday/YouGov research, only 9% of people said they had difficulty believing the theory of evolution—the vague sense that it is incompatible with religious belief persists. Twenty-two per cent of Britons disagree that 'it is possible to believe in God and in evolution', compared to 48% who agree, and the balance in the US is more sceptical still. Around 40% of Americans, overwhelmingly religious ones, disagree that humans 'developed from earlier species of animals',[2] and, in one 2020 study of biology students, researchers found that over half (57%) 'perceived that evolution is atheistic even when they were given the option to choose an agnostic perception.'[3] In short, however science might be formally defined, it is popularly understood to be characterised by various 'beliefs' about the world and the universe, and some of these beliefs—particularly those around cosmology and evolution—are perceived by more than a small minority to be sources of tension with religion.

Crucially, however, it is the contention of this chapter that the real heart of this particular conflict over content is less about cosmology and evolution per se and more about anthropology that is believed and felt to be implied by them. It is about how those disciplines sometimes generate an understanding of the human that is judged unacceptable to religious doctrine (and vice versa).

To be clear, we want to avoid overclaim here, and do not intend to force *all* content-based tension into this procrustean anthropological bed. Sometimes the disagreement is straightforwardly based on the conviction that a particular scientific idea, say about the origin of the universe, is incompatible with a particular religious belief. However, our interviews strongly suggested that, time and again, when people disagree about the content of science and religion, they are disagreeing about the implications for our understanding of human beings.

198 THE LANDSCAPES OF SCIENCE AND RELIGION

If so, this would concur with both John Evans' and Elaine Howard Ecklund's analysis of *popular* tensions in the science and religion debate in the US. Evans, as we noted in part I, argues that one of the pinch points for science and religion dialogue in popular discourse concerns the 'implicit moral ideas [that] are embedded in scientific claims, and particularly those scientific claims, like Darwinism, that have implications for the nature of the human'.[4] In a similar way, Ecklund locates religious America's widespread problem with Darwinism in two key issues that often lie submerged beneath the surface of the debate, one of which is what evolution means 'for the sacredness of humanity'.[5] The research in this chapter supports and expands those conclusions.

The chapter begins by looking at what people are disagreeing about in the science and religion landscapes when it comes to the content of the physical sciences, specifically cosmology. It then moves on to the same issue with the life sciences, specifically but not exclusively evolution. In the process it draws out the argument of how the nature and status of the human lies at the heart of both these objections. The following section goes on to explore in greater detail what's really at stake here. What exactly are people disagreeing about when they are disagreeing about human nature in the light of the science and religion conversation?

The chapter then proceeds, in the fashion of previous ones, to look at how the gaps we have just explored can be closed up. It is our opinion that the tensions between science and religion in this area, despite having wider public purchase than those described in previous chapters, are easier to settle. There is disagreement here, but it is not as acute as, say, the dispute over metaphysics. That doesn't mean the dispute has been fully settled, however, and the final part of the chapter explores what we consider to be some outstanding questions concerning the apparent anthropological tensions between science and religion.

Cosmology

Despite the fact that, in the quantitative research, cosmology (and the Big Bang) were seen as a bigger threat to religious belief than that presented by other disciplines, public opinion was still relatively balanced on the issue, with roughly the same proportion of people agreeing and disagreeing that cosmology made religious belief hard.

This balance was also evident among the expert interviewees, who voiced opinions 'for' and 'against' in roughly equal measure. Although some cited the fact that the scientific account of 'creation' disagreed with that found in religious texts, few thought this alone was a serious point of tension.[6] A creative, contextualised, literary reading of those texts, most obviously Genesis, was commonly recognised to be far more authentic and legitimate. Nobody played the 'creationism is the only authentic Christianity' card.

More substantive was the view that the scientific theory of the universe's origins rendered any 'God hypothesis' redundant (#1). Invoking a God 'who lit the touch paper' was 'ultra-naïve' and 'doesn't help' (#33). The idea of creation *ex nihilo* was fundamentally 'unintelligible' (#5). Any idea of a creator simply invited an 'infinite regress' which answered nothing (#46).

Against this, some observed that the very fact that the universe had an origin at all, as opposed to being eternal as had previously been thought, was at least conducive to the idea of a first-cause creator and a challenge to any wholly naturalistic explanation of existence.[7] Moreover, it wasn't simply that the universe had an origin but that it had an apparent order, a 'fine-tuning' that was both extremely delicate, exceptionally unlikely, and highly conducive to life. 'The physicists tell us that it is absolutely incredibly improbable that that would happen by chance. And so, in a fairly straightforward way, that does give us evidence for design of some kind' (#17).

Predictably, opinions were divided on this 'anthropic principle' or 'Goldilocks hypothesis', although even those strongly ill-disposed to any religious interpretation acknowledged that such fine-tuning did at least pose a question.

> Some theists get very excited about it and think that there's some evidence here of design, some kind of teleological argument can be run and maybe it could. Maybe that should tip us a little bit in the direction of some kind of intelligence behind the universe, possibly. However, in my estimation, even if it did do that a tiny bit, it would only be a tiny bit. (#42)

To counter this theistically flavoured interpretation, critics pushed back with three possible explanations. One was simply luck—there was no a priori reason why it couldn't be chance. The second was 'quantum fluctuations' in a 'vacuum'.[8] The third was the idea of the multiverse, that our universe was only one of many, meaning we need not be so astonished by the universe's fine-tuning or invoke some kind of divine intelligence to explain it.[9]

200 THE LANDSCAPES OF SCIENCE AND RELIGION

Predictably, again, there were responses to these responses. Invoking 'luck' was no real, certainly no reasoned, answer at all. Quantum fluctuations was not so much 'a creation story [as] ... a creative accounting story' (#12), a classic example of the 'one free miracle' defence.[10] 'You can get away with murder in quantum mechanics because people think they don't understand the equations so they can't judge the metaphysics' (#12). And the multiverse was hampered by being 'empirically meaningless' (#7), 'a get-out-of-jail-free card' (#47), not only without 'a shred of evidence' in support of it (#41), but also literally untestable, as well as flying in the face of Occam's razor by positing 'an infinite number of universes [just] to get you out of the idea of [one] God' (#43).

There were further responses to these responses,[11] but by this point it will be clear that however absorbing these conversations about religion and cosmology were, they clearly weren't tipping the overall balance strongly in one direction or the other. Moreover, in an odd reversal of received wisdom in all this, the theologians to whom we spoke were distinctly lukewarm about the alleged religious implication of this whole field, whereas the cosmologists (and indeed other scientists) were often of the opinion that theirs was one of the more spiritually inclined scientific disciplines. 'I've never been convinced by the cosmological argument for the existence of God in the first place', claimed one theologian (#6),[12] while one science journalist remarked how 'a lot of cosmologists find themselves attracted to the idea that there is a higher power that somehow created the universe' (#13).[13] Ultimately, however fascinating the cosmological debate was, and however much people did disagree about it, this was not really a conflict *between* science and religion so much as a series of disagreements within science and within religion.

Where this changed was in the subtler, unspoken implications that the whole cosmological conversation had on our understanding of human significance. Disagreements about the origin and nature of the universe—about quantum vacuums, fine-tuning, multiverses, and the like—were ultimately ambiguous and unsettled, but what they did appear to do was to put humans into perspective, rendering them almost invisibly small, insignificant, and, crucially, not the kind of creatures envisaged by religious doctrine. Although the idea that Copernicus's heliocentrism which demoted and demoralised humans owes far more to Freud than it does to sixteenth-century intellectual history,[14] one philosopher nonetheless cited this example to explain what he meant:

I don't think that the importance of Copernicus, that the Earth isn't the centre of the universe, that we're one little planet in a huge possibly infinite space, I don't think the significance of that has gone away. So I think all these discoveries have made the position of traditional religion difficult in a cumulative way. (#55)

This view came in a number of forms. One interviewee elided the idea that humans were made in the image of God with the idea that they 'are in the centre of it all' (#17), an elision that is also implicit in the quotation above. In this way, losing centrality necessarily means losing significance.

Others made the same point with reference to time rather than space. Humanity was temporary,[15] as indeed was the universe itself. '[I am] troubled by the fact that all our physical theories suggest to us that in one way or another, the universe will end with lifelessness and perhaps with a big rip at the end of the universe itself' (#10). Admittedly, those in the know observed that the timescales here might be considered to be rather too large to worry about.[16] But the point remained that the stage on which human existence was being played out appeared to undermine the significance accorded to it by religious doctrine. 'The idea of a God who gathers up all our purposes into his or its or her overriding purpose. It doesn't make sense. It just doesn't. Purposes are irreducibly various, and irreducibly transient' (#12). Cosmology was a threat to religion because it undermined humanity.

The same point was made, albeit more colourfully, by the walk-on role that alien life and astrobiology played in discussions. The discovery of alien life, let alone alien intelligence, would pretty much confirm the idea that the universe was made 'for' life.[17] This could be deemed to support the religious point of view.[18] However, for a number of interviewees, such a discovery only further undermined the theological concept of the human. 'The discovery of life anywhere else', we were told, would 'undermine the human-centredness of religious writing' (#10). It would certainly pose interesting questions.

If we came across another intelligent species from another part of the galaxy, what God would they believe in? Where does God fit into that? Would they have religious deities? Would these be great people? Would they have souls? Would we recognise them, allow them into heaven? (#39)

While none of these were considered to be unanswerable (and indeed some of them were answered), they further contributed to the idea that in as far as the content of cosmology posed a threat to religious doctrine, it was most acute around the theological concept of the human.

202 THE LANDSCAPES OF SCIENCE AND RELIGION

Evolution

The apparent tension between evolution by natural selection and religious belief is firmly established, with innumerable witnesses from very different quarters providing evidence in support of a conflict. Indeed—magnified by some of the court cases mentioned in part I; by the growth of ID and fears about science education in and beyond the US; by iconic books and interventions from leading atheist scientists insisting that Darwinism does away with the need for God; and by reams of public opinion research, which shows that, contrary to some predictions,[19] this is no longer simply a localised American and Christian fundamentalist phenomenon, but one that is growing widely, especially in the Islamic world—the apparent tension here has arguably become *the* science and religion battleground over recent years.

Our expert interviewees were less convinced about the centrality of this conflict. They cited various counter examples. The Catholic Church had long since accepted evolution. The panic over faith schools was a bit of press hysteria.[20] Playing the (anti-)creationist card was a simply useful way for aspiring biologists to 'break into the media'.[21] Much of the debate was mere culture war ammunition.[22] And only fundamentalists got worked up about Genesis, and sometimes not even them. 'Even in the deep south in America, where fundamentalists are strong, in the 80s and 90s', one sociologist told us, 'I really didn't find that many people who were bothered about Genesis 1-12' (#22). Ultimately, in the words of one science writer, 'I see no thoughtful, serious-minded religious leaders having any issue with the standard view of evolution' (#10).

However valid these counter examples, the fact remains that many people *do* perceive a tension in this area. As already noted, only a small minority (9%) of people in the UK say they have difficulty believing in the theory of evolution (though, as we have also seen, more think evolution and religious belief are in conflict, they simply reject the latter rather than the former). When we asked this 9% why they had difficulty, offering them a set of pre-selected answers, it became clear that the problems clustered around the issue of what evolution meant for human beings. Thus, while only a few people had difficulty in believing that animals (15%) or plants (12%) evolved, rather more had problems believing that humans evolved (34%) or believing that 'humans and apes share a common ancestor' (36%), or that 'all life, including humans, has a common origin' (28%).

In a similar vein, while 74% of the total sample agreed that 'there is strong, reliable evidence to support the theory of evolution', this rose to 80% when people were asked whether they thought that 'plants and animals have developed over time from simpler life forms' but crashed to 28% when it came to whether evolution could account for human consciousness[23] or morality.[24] This is strongly in line with the research conducted by Elaine Ecklund, and points towards a pattern that was amplified in the expert interviews.

To be clear, by no means did all of the expert interviewees claim that evolution was in clear conflict with the content of religious belief. If anything, this was a minority opinion, even among the atheist scientists and philosophers we spoke to. Moreover, when expert interviewees did locate part of the science–religion tension here, it was not *necessarily* connected with the question of the human. Sometimes, the tension was simply that evolution apparently demolished any theological arguments from design[25] or that it was sufficient and crowded out the need for any other explanations.[26]

That recognised, the core of the perceived tension did pertain to human beings or, more precisely, to the idea that 'creation' was the kind of thing that was in any way interested in or supportive of human values and concerns. Evolution revealed a world and a past that was wasteful and painful on an unimaginable scale.[27] It was a process that was apparently without any guidance, with no supervisory, let alone interventionist, role as envisaged by some theories of theistic evolution.[28] It was governed by chance,[29] interrupted by catastrophes,[30] and resulted not in exquisite design but in 'a mess'.[31]

Once again, it is important to be clear here. None of these arguments claimed to disprove the existence of God or to refute any dogmas that were deemed central to religion (except the Paleyian argument for design, the significance and centrality of which was long past). It was hardly as if religious thought was indifferent to questions of waste, pain, chance, and suffering. Indeed, as one interviewee pointed out, religious belief is more likely to be borne from the protest against such problems than from any satisfied acceptance of nature's harmony and perfection.

> A sense of order and design is not what leads most people to theological conclusions. Indeed, it's anguish and disorder. It's the interplay between the intuition that the universe ought to make sense and the fact that it doesn't which drives religion. I don't think you can get rid of it by claiming we've resolved it with natural selection. You have resolved Paley's argument. Great. Christianity lasted for eighteen hundred years before Paley. (#2)

204   THE LANDSCAPES OF SCIENCE AND RELIGION

Nor was it the case that these arguments were in themselves unanswerable. As we shall see, there were plenty of arguments deployed against them.

Rather, the issue here lay in the feeling—sometimes it was no more precise than that—that the world as revealed and understood by evolution by natural selection simply did not correspond with that envisaged by religious doctrine. Creation was cruel not good. Events were random not planned. (Pre)history was directionless not designed. And, most acutely, humans were accidents, and emphatically not the kind of creatures envisaged by religious doctrine or salvation history.

Anthropology

What exactly was the problem here? If the humans revealed by evolution, and cosmology, and, as we shall see, psychology and neuroscience were not the kind of creatures envisaged by religious belief or doctrine, why were they not? Where precisely did the tension lie?

Eight overlapping but distinct points emerged in our discussions. Some were deemed more serious than others and none was judged decisive. Indeed, it is important to emphasise, as we shall see, that the pushback against these objections could be as forceful as the objections themselves. However persuasive that pushback was though, it was unarguable that science (primarily in the form of evolutionary biology and psychology, but also through neuroscience and, as we have seen, cosmology) is perceived to present a challenge to religious belief and doctrine vis-à-vis its understanding of human existence, worth, dignity, and purpose.

First, science undermines the idea of the immortal soul. At its most basic, an evolutionary account of human beings challenges any notion of substance dualism when it comes to the person. As we noted in the chapter on metaphysics, this is part of a wider scientific challenge against all non-naturalistic accounts of reality, but it attained a particular force when it came to religious conceptions of the human, a conception for which soul-language is so important. In the words of one professor of evolution, 'as we learn more about the workings of the brain and consciousness then, yes, I think there is less and less room for a soul to exist as a separate entity' (#72). Or, in the more direct words of a philosopher, 'the immortal soul is a completely anti-scientific concept' (#3).

This challenge was deemed an empirical issue rather than simply a metaphysical one because, it was felt, some scientific disciplines had come close to proving the negative, peering into the human and finding no evidence of what religious believers claimed to find there.

> You can look at some of the classic studies of people who have brain injuries, and have parts of their brain knocked out, and completely change their personalities, lose certain aspects of morality, or indeed behaviour and moral sense. I think it is hard to cling onto the idea of this sort of inviolability of the soul. (#39)

As this quotation makes clear, that empirical challenge is directed not simply at the soul but at human moral sense. This is the second of the anthropological challenges to religion: evolution explains morality.

According to this view, human morality is essentially 'an extension of inherent adaptive behaviour' (#9). 'We have the moral beliefs we do because they have been good for survival, not because they track some independent moral facts about right or wrong' (#17). 'Evolutionary game theory' explains moral concepts like 'reputation, direct reciprocity, indirect reciprocity and so on' (#50). Ethics is simply 'an emergent property of how we socially interact and transmit that consciousness around' (#27). 'Kin selection and reciprocal altruism' are the cause of our 'hatred, love, friendship, sympathy, empathy, compassion ... [the] emotional things upon which morals rest' (#24). Few claimed that this was a challenge only to religious belief, but the close tie between religion and morality—both at a personal and at a more formal, codified, doctrinal level—meant that it was felt to be a particularly acute problem for religion.

Part of this moral challenge lay in the fact that this evolutionary approach to ethics left little that was distinctive to humans. 'All social animals, by definition, have to have cooperation from the social group to survive, it's not just humans. All social animals have some, what you could call, ethics ... it's not formulated in words, but they have a cooperative principal' (#41). And this led to the third challenge: to human uniqueness.

Whatever else religious groups—usually among our interviewees, a cipher for those within the Abrahamic tradition—believed about human beings, there was the conviction that they were somehow special, distinct and different from other animals. Evolution suggested otherwise. Humans, we were told, cannot now be considered as 'separate' from other animals.[32] They are not 'special'.[33] 'Most churches seem to have come to terms with the fact that we are evolved creatures [but] how they square that with the special space

206 THE LANDSCAPES OF SCIENCE AND RELIGION

of mankind in creation is not always clear' (#12). Evolution did the heavy lifting here, but it was not alone, and neuroscience played an ancillary role in this objection.

> I think religion still has this notion that humans are something special, a special creation of God and that that sort of sets us apart from the animals and there's [a] lot of really interesting traditions where humans have dominion over animals. Neuroscience has demolished that. (#13)

Challenging human uniqueness easily morphed into a fourth challenge, closely related but distinct in its precision, namely to the idea that humans were made 'in the image of God'. Time and again we heard that the evolutionary history of mankind precluded our bearing the *imago dei*. This view was more often assumed and stated than explained and defended, and when it was defended it could be done in what seemed like a rather wooden, literalist way. 'We are biological organisms, we are sexual organisms, we are mammals, these things are not incidental to what we are. God's not a mammal or a sexual being' (#3). Nevertheless, there was a palpable feeling—again, it was sometimes no more specific than that—that the messy contingency of evolution was incompatible with the dignity implied in bearing God's image.

A fifth challenge came from the perceived threat to free will. This took a similar form to the threat to morality, partly in the way it came from both evolution and neuroscience, and partly in the way that, however much it presented a generic threat to the ideas of the human, it was felt to be a particular threat to the religious conception. The evolutionary threat to religious free will was oblique and came primarily through evolutionary psychology. Evolutionary psychology was now able to provide robust explanations for human belief and behaviour, and in doing so it undercut not only religious explanations but also those common-sense explanations based on will or freedom. Evolutionary psychology generated 'potent explanations for why humans are altruistic, why they love, why they care about their children, why they care about their families, why we behave in ways we call—in inverted commas—virtuous' (#39). In doing so it undermined the idea, closely associated with religion, that 'I freely chose x belief or y course of behaviour' was in any way an explanation.

A more direct a challenge to free will came from neuroscience. In the words of one science journalist, 'neuroscience has eroded and almost demolished the idea that we have free will', and, in the words of another, 'if we didn't have a degree of free will, all religion would be undermined' (#54). In reality, this was not much different to the older, though for some still live, challenge to free will from a causally closed, physical universe. 'I ... remember talking to 150 physicists once and discovering that 149½ of them didn't believe in free will, but nevertheless, I think the Christian theologian can't part company with the reality of freedom of will' (#100).[34] Neuroscience brought the old problem of closed physical causality to bear on the presumed seat of human will.

A sixth challenge to religious belief was the idea of religious experience. If some combination of evolution and neuroscience could undermine the morality and free will that was so fundamental to religious anthropology, they could do the same to religious experience. Several interviewees mentioned Michael Persinger's 'God Helmet' as an example of how religious experience could be induced and was therefore effectively nothing more than 'electrical activity in the brain' (#23). Religious ecstasy might just be unusual brain activity, and the religious titans of the past just possessed of 'unusual brains'.

> The more and more we know about the way the brain works, the more and more we can see how ... certain religious figures in the past have had unusual brains, maybe have had temporal lobe epilepsy, for example, that's caused them to have visions that become the basis of faiths at a later date. (#72)

It was a short step from offering scientific evidence that explained (away) religious experience to having scientific evidence that explained away religion itself, the seventh challenge. Here the emphasis shifted back to evolution but also to anthropology. 'Instead of seeing religion descending from heaven', as one philosopher put it, the scientific account of religion revealed it to be little more than 'our attempt to realise our transcendence' (#12).

Various possible reasons presented themselves. Some were long-established. Religion originated as a kind of 'protoscience'.[35] It was 'a way of explaining things for people'.[36] Or it was the consequence of our unique cognisance of our own mortality—the apparent uniqueness of which rather ran against the objection that there was nothing unique about humans—and was therefore simply a way of warding off fears of death.[37]

More recent and fashionable explanations were also proffered. Religion was essentially a 'cognitive by-product' of evolution.[38] It was the result of HADD, which explained why 'we're so very prone to believing, holding false positive beliefs in some sort of invisible agency, be it a ghost or a dead

208 THE LANDSCAPES OF SCIENCE AND RELIGION

ancestor or a spirit or a God, or whatever it might happen to be' (#42). Or it was a consequence of group size in our recent evolutionary past.

> What essentially humans have done is find a lot of behaviours which trigger the same underlying neuropeptide system, the endorphin system in the brain, which is the foundation of bonding in primates, without having to physically touch another individual. And this is where the rituals of religion come into play as nearly all of them seem to be very good at triggering the endorphin system ... trance processes and those experiences that you have in that sort of internal world are a consequence of the endorphin system or underpinned by the endorphins because they are opiates, basically. You can get an opiate high from them. (#51)

It is interesting to note that these various scientific 'explanations' for religious belief and behaviour reflect and map onto the range of understandings or 'features' of religion that we drew out in part I. Evolution and neuroscience can, it is claimed, explain the cognitive, communal, ethical, and ritual features. They can explain why religions falsely believe in gods and spirits, why belief in a post-mortem existence is so important to them, why religions posit a causal role for the divine, why religion has such a profound community element to it, and why religious ritual is so important. Whatever one's conception of what religion is, there is an evolutionary, anthropological, or psychological way of explaining it (away). In reality, however, what matters here is less the content of the explanations than the simple fact that they undermine the idea that religion is about what it thinks it's about. Religious believers think they are tending to fate of their souls, or communing with the divine, or building the kingdom of God, or following Jesus, or obeying God's law, or submitting to Allah, but they're actually just managing a form of sophisticated, large-scale cooperation or clinging to an outmoded explanatory system.

The eighth and final challenge is to any kind of salvation narrative that posits a particular concern for humans. At first, this appears to be a problem just for the Abrahamic faiths, but, in reality, it is a challenge to any religious belief that conceives of humans achieving any form of transformation that is not also available to other animals. And it is a challenge because evolution insists there is no line in prehistory between 'humans', however they are defined, and what preceded them.

This was the point at which Neanderthals played their significant walk-on role in our discussions. A surprising number of interviewees, not all of them anthropologists, asked whether Neanderthals (or 'cave men' or 'protohumans') had souls, or whether they could be saved, or could 'go to heaven', or were loved by God in the same way as members of our species were (#13, #47, #40, #52, #79). It was a similar kind of question as the one that was posed by the possibility of intelligent aliens. Why are humans special in salvation history?

The very heart of the problem lay less with any event itself, such as the point at which humans first became humans, or recognised the transcendent, or were granted souls, than with the very challenge of gradualism. 'I don't think religion's very good at very gradual evolution of things like that. It tends to work on, this happened then, at a point, rather than it gradually happened' (#47). Evolution showed humans as part of a continuous prehistory with other hominids, and this seemed to erase any hard lines that religions wanted to draw around humanity.

In summary, then, we identified eight challenges that science put to religion when it came to the issue of their respective 'content'. The disciplines of evolution, psychology, anthropology, neuroscience, and, more distantly, cosmology generated a certain content—a series of ideas about the nature of life, the universe, and, above all, humanity—that was in apparent tension with many religious beliefs. The challenge was to the (immortal) soul, to morality, to human uniqueness, to the idea that humans were made in the image of God, to free will, to religious experience, to the origins and exercise of religion itself, and to any doctrine or salvation narrative that drew a firm line between humans and other hominids.

All these were judged as threats to religious doctrine and belief, but not universally or unanswerably so. These objections were, to repeat ourselves, neither unanimously held nor (usually) considered to be knock-down arguments. Accordingly, we turn now to the different ways in which people, and not just the religious experts to whom we spoke, pushed back along this complex front.

The pushback

The pushback—the argument that there was much less or indeed no tension at these points—took various forms. They can helpfully be categorised into three groups, the first pertaining to science, the second to religion, and the third, more conceptually, to the very nature of the conflict in this area. We will look at each in turn.

210 THE LANDSCAPES OF SCIENCE AND RELIGION

Revisiting the science front

Interviewees critiqued specific scientific experiments, conclusions, and sometimes entire disciplines for their role in this part of the alleged conflict. The idea behind Persinger's God Helmet and, more generally, the idea of using magnetic resonance imaging (MRI) scans to explain away religious experiences was dismissed. Quite apart from anything else, MRI scans were a helplessly blunt instrument for this kind of work.

> Functional MRI and all the other techniques ... it's just, to put it at its crudest, an extremely expensive hi-tech version of phrenology because you're looking at such large structures in the brain and the brain is so complex and the microstructures are so elaborate that we're just looking at big lumps. (#61)

Perhaps more pertinently, the conclusions drawn from them were a complete non-sequitur.

> If we go into Andrew Newberg's research and we see that you have a deeply pious Buddhist monk who is meditating and having a particular sort of neural experience and this is identifiable and empirically capturable with MRI or some other brain imaging technology, what have we done? What have we discovered? I think at most we've discovered that something is happening in the brain when people experience God. (#23[39])

Such experiments proved nothing more than that the brain is involved in religious activities, just as it is in all human activities. As one astronomer put it, 'just because you may be able to trace these things to neurons firing ... does not negate the reality of the experience' (#8).

A similar objection was raised concerning the neuroscientific claims against free will. Several interviewees cited Benjamin Libet's experiments apparently disproving free will as an example of bad science and false reasoning. It was 'an extremely simplistic way of thinking about free will' to imagine that 'a decision having been made in your brain before you are consciously aware of it' somehow invalidated free will. 'It seems to me that that's a regular part of the human experience' (#10).[40]

The gene-centred, neo-Darwinian interpretation of evolution was also questioned. Even if this theory did make it hard 'to accommodate religious belief', there were other, more amenable understandings of evolution. The 'niche construction theory' and the 'extended evolution synthesis theories', for example, allowed for a greater role of organisms themselves in the process of their evolution. Instead of 'just being a passive template on which different traits are selected for, according to Darwin's theory of natural selection', these approaches allowed for much more interaction between organisms and the world in which they lived. 'It is a much more dynamic system ... where, to a degree, the creatures that are evolving also have a say, to an extent, in their own evolution by the actions that they take' (#82). Such an understanding of evolution, it was argued, was more agreeable to the religious worldview to which Neo-Darwinism was apparently so inimical.[41] 'Epigenetics' and the idea of the inheritance of acquired characteristics, until recently a heresy in evolutionary circles, was also cited.[42]

Sometimes it was not the theory that was questioned so much as the conclusions drawn from it. Again, evolution was in the firing line here. A number of interviewees strongly contested the idea that evolution by natural selection was wholly arbitrary and random. Evolution was constricted. 'There is a structure in the world within which evolution is constrained' (#82). It wasn't simply contingent. 'History is blind but that doesn't necessarily make it contingent. It's still constrained by principles of selection and the operation of populations and groups and those sorts of things ... Contingency is a cop out really. It's a non-explanation' (#52). Evolution was also convergent, repeatedly homing in on familiar solutions to various problems.[43] None of this meant that it was teleological, but perhaps 'there is a sense in which there is a direction to [it]', a sense 'that is more compatible with theology than the previous version, which tended to deny any kind of sense of purpose at all' (#82).

In a similar way, the natural explanation for religion in human prehistory, as revealed (or at least posited) by evolution and anthropology, was defused or even turned to religious advantage. The religious urge appeared to be deep and hard-wired into human nature. Anthropologists mentioned cave painting, burial traits, and jewellery. 'Pro-social ritualistic behaviours are natural to us, are engrained in our biology' (#83). Our 'special attention to the dead' (#66) goes back possibly hundreds of thousands of years.[44] Prehistoric cave art is 'part of a belief system, a complex belief system, what we might call cosmological religious belief system' (#72). We are *naturally* 'introspective' and cognisant of our 'mortality' (#66). We cannot but use our imagination, 'our inherent sociality ... gives rise to imagined beings' (#63[45]). We are 'narrative seeking'

212 THE LANDSCAPES OF SCIENCE AND RELIGION

(#61) creatures, inalienably meaning-*seeing*[46] (and meaning-seeking), and relational at our very deepest level. 'We're not born alone, we don't exist outside of relationship, we're always inside a relationship with each other' (#71).

Humans might not necessarily be unique in all this. One expert in the subject posited that while there is no evidence that Neanderthals had any religion, it was not beyond the realm of possibility that they had 'a transcendent engagement with the world'. They did, after all, have an aesthetic and possibly symbolic understanding (#79). More daring still, one archaeologist spoke of the 'effervescence'—a loaded term in the discussion of religion—that was present 'when elephants meet after separation', something that is 'a feature of social animals' (#52).

However far back such traits could be traced and however much they were not uniquely human was largely beside the point. Indeed, the further back they could be traced, arguably the more they strengthened the religious pushback. The critical point was that the more evolution, archaeology, and anthropology revealed of the natural history of religion, the more religion seemed to be natural. Humans were naturally relational, moral, existential, symbolic, introspective, imaginative, transcendent beings, which was a roundabout way of saying they were naturally religious beings. In the words of one non-religious sociologist, this rather turned the tables on the idea of religion as something that needed to be explained.

> There is such a strong sense of the way in which humans become religious that it has left that question of how the heck do people become non-religious. The interpretation that it is purely the product of the human brain isn't one I have encountered as much as the idea that the human brain is wonderfully designed to receive and understand God.' (#30)

This could be put more polemically, as one theologian did by referencing the work of Justin Barrett, an experimental psychologist who has conducted a great deal of work on the natural belief in God.

> If there were a God and God desired to be in a relationship with humans and the rest of creation, wouldn't you expect that humans would have the capabilities and natural dispositions that would then allow them to have the relationship with God that God intended? … You wouldn't expect religious beliefs to be contrary to everything that humans are naturally wired or predisposed to do and be. (#23)

ANTHROPOLOGY 213

In this way, scientific ideas along this front could be doubted, questioned, revised, or even altogether transformed. And occasionally they were demolished. Undoubtedly the most brutal critique of the role that scientific disciplines played in this apparent area of conflict was directed at evolutionary psychology, a critique that came not from wounded theologians but primarily from biologists and psychologists themselves.

The idea that 'everything can be explained through what humans may or may not have been doing 50,000 years ago' was not persuasive (#22). Evolutionary psychology was 'about manipulating whatever set of ideas you have about the world in order to suit your politics' (#48). The discipline was full of 'just so stories' (#35, #36, #37, #39), 'post-hoc rationalisations' (#37), 'low-quality research', some of it 'trying to explain religious ideas as having evolved' (#39). One eminent biologist engaged in an extended complaint against the discipline, which was simultaneously both informed and amusing.

> [It] undermine[s] the human condition by applying really bad science and evolutionary reason ... I'm going to give deliberately crass examples, but they are from real scientific papers. So, if you say that why women wearing blusher or rouge is an evolved mechanism that reminds men of fruit ... or babies cry at night to prevent their parents from having sex and therefore creating more babies which compete with them, all in a peer-reviewed academic study ... Women prefer pinks and reds because in the hunter/gatherer phase of our existence berries are red ... almost all of these examples fall apart at the merest degree of scrutiny. (#35)

The *coup de grace* was savage, the interviewee deploying perhaps the most damaging comparison possible. 'You get an almost religious adherence to these ideas ... evolutionary psychology is effectively sort of similar to creationism'.

This, then, was the first part of the pushback against the idea of a fundamental conflict here: the scientific front was not as firm as some liked to think.

Revisiting the religion front

The second part of the pushback came along the religious front, specifically in re-examining the beliefs, ideas, and doctrines against which these scientific disciplines were apparently pushing.

214 THE LANDSCAPES OF SCIENCE AND RELIGION

Take the (allegedly) religious idea that humans were unique. Whether through the work of zoology, evolutionary anthropology, or cosmological speculation, this was eroded, thereby, at least according to those who located a conflict here, leaving an important element of religious doctrine in tatters. This argument itself was critiqued on both empirical and theoretical grounds. Received wisdom might like to claim that there was nothing special about humans, but plenty of interviewees pointed out that this simply wasn't true. Our level of intelligence, our use of language, our imagination, our capacity for moral evaluation, our awareness of mortality, our religious impulse, our ability to modify our environment and therefore exercise disproportionate control over the process of evolution ... all of these were unique to us, at least to the best of our current knowledge. And, unless and until we encountered intelligent aliens who showed similar traits, or found certain earth-bound animals that were vastly more intelligent, morally reflective, religious, etc. than we heretofore thought, we would remain unique.

In any case, the very concept of uniqueness itself was a red herring. Every species was unique, every species was 'special'. Indeed, that is pretty much what the word species means, species being to genus what special is to general. 'There is a uniqueness to every species' (#69). All that noted, those theologians and believers who addressed this issue were at pains to point out that the 'scientific' understanding of the religious conception of specialness was, in any case, way off the mark. However 'special' religious doctrine might consider humans to be, it hardly denied that we were still a material part of creation, sharing much in common with other creatures. This applied to even the allegedly most anthropocentric of religious traditions. 'If you read the Genesis account of creation, there is nothing special about human beings. They're exactly the same as creatures, they're filled with the same spirit in their flesh. They are certainly part of creation' (#34).

Moreover, according to Jewish, Christian, and Islamic thought, human 'specialness' resided not in our inherent or innate characteristics but in the grace of God. Theologically speaking, human specialness might have been *enabled* by our particular (imaginative, moral, cognitive) capacities, but it was grounded in the fact that we had been chosen, or spoken to, or revealed to, or incarnated amidst by God. 'In the Hebrew Bible, and in Jewish tradition, you get much more of an emphasis on God elevating human beings to that status' (#64).[47] Human specialness was not a result of merit or capacity or distinctness but of grace and love.

This confusion about human uniqueness was closely linked to that about human worth and dignity. As we saw above, there was a belief that what science had revealed about life and the universe undermined any religious conception of particular human worth. Just as humans weren't unique in their capacities, nor were they uniquely valuable or worthy of the attention lavished on them in religious doctrine. And yet, as a number of interviewees pointed out, this was pretty much the same error as regard to human uniqueness. 'My worth as a human being is not because I am different from the rest of creation. My human worth is because I am loved whatever physical or mental makeup I happen to have' (#16). Significance and meaning were not rooted in size or position but relationship. In the words of one cosmologist:

> The size of us and the size of the universe is just completely, awesomely different. But biblically our meaning is not in our position ... our meaning isn't in that the earth is the centre of everything. Our meaning is that God loves us. And that's very different ... we don't take our significance from being at the centre of the universe ... we don't take our significance as there being only one earth ... We take our significance from our relationship with the one who made it. (#43)

There was a third confusion around the apparent challenge from the content of science, whether neuroscience or anthropology, and this was to the religious idea of the soul. Here the pushback was ambiguous. On the one hand, it was admitted that many religious believers and some theologians did adhere to a 'substance dualist' conception of the soul, possibly implanted into the human species at some point during its evolutionary past and into the human body at some point after conception. This, as we have noted, did not sit easily with the scientific understanding of the human.

However, others dismissed this view as fundamentally alien to (at least the biblical) religious worldview.[48] The idea of a separate soul 'is a Greek imposition again upon the Bible' and 'has nothing to do with either the Old Testament or the New Testament', which understand God as 'the one who interacts with us as whole human beings that is body and mind, spirit, soul altogether indistinguishable or undividable' (#16). Over and against the dualistic or Platonic conception of the body and soul, it was, in the words of one theologian, 'quite plausible and quite consistent ... to think of humans in a physicalist way'. Christian thinkers like Nancy Murphy and Warren Brown, he explained, would be 'very sceptical about the language of the soul', or at

216 THE LANDSCAPES OF SCIENCE AND RELIGION

least sceptical about the language of a *separate* soul. Rather, 'if that language has any use we should understand the soul as an aspect of the human being, not a part of the human being' (#97). This was a view more consistent with an Aristotelian conception of the soul, or an updated Aristotelian conception such as was held by theologian–physicist John Polkinghorne, which conceives of the soul 'as being the almost infinitely complex, dynamic, information-bearing pattern in which the matter of our bodies at any one time is organized.'[49]

In summary, it frequently seems that the religious front against which certain scientific disciplines were pushing here was, in fact, a fake front. Biology showed that humans aren't unique, but religious belief wasn't dependent on them being so (at least in the way critics claimed). Cosmology showed that humans were insignificantly small within the universe, but that did nothing to undermine the religious commitment to human worth and dignity, which was never grounded in size or centrality in the first place. Evolution and neuroscience strongly suggested that humans did not have an immaterial soul, but the kind of soul that these disciplines disproved bore little resemblance to the way in which (many) theologians and believers conceived of it. Time and again, it emerged that the beliefs that MRI scans or evolutionary psychology or astrobiology have slain looked nothing like those held by religious doctrine.

Revisiting the whole encounter

So, just as some of the scientific disciplines, theories, and interpretations that pressed against religious doctrine could be less threatening than imagined, so some of the religious doctrines that were thought to be under threat turn out to be not particularly religious at all. Both of these factors revised the battle lines that were apparently drawn here and, in doing this, they were joined and aided by a third argument, which, in its own way, constituted an altogether more significant problem to the very idea of there being conflict between science and religion concerning the content of both.

This emerged through the repeated observation that many, perhaps most, of the apparent threats posed to religious doctrine by evolution, anthropology, and neuroscience were not limited to religious doctrine. On the contrary, these disciplines challenged and undermined our familiar and fundamental understanding of the human, an understanding that is common to a wide range of beliefs, religious and not.

For example, the same line of scientific attack deployed against 'the soul' could be and was used against 'the self'. The 'scientific view of the world ... reduces the human self to nothing effectively' (#54). In as far as it permits the self to exist, it is as 'a construction, it's a story, it's a model, a representation of a "self" that doesn't in any other sense, exist' (#24). More damning still, the self is no more than an evolutionary con-trick. '[The] self is an illusion created by evolution, right? I mean, otherwise you wouldn't be very efficient at evolving ... scientists who are religious might want to reflect on the fact that evolution has faked up your neurochemistry' (#27).

In a similar vein, the threat posed by evolutionary psychology and neuroscience to religious morality was really just a subset of a much broader threat to morality, to any form of moral realism, and to the idea that humans were capable of reasoned moral evaluation. It was the same with free will, the denial of which affected all human behaviour.[50] Consciousness was also in the firing line—'there isn't anything more to being conscious than being a physical system'—although those who put forward this view were often vigorous in denying that they had denied the reality of consciousness.[51] In short, most of what constitutes human experience can be explained away.

> If you take seriously the scientific viewpoint ... [you end up with] this idea of the world being completely indifferent. There is no values, meaning, purpose or anything ... You follow through all these things, and it is hard to make sense of ourselves as human beings ... I am not sure many scientists really do. (#14)

Virtually no-one we spoke to was willing to countersign this understanding of the human. 'I refuse to accept that the glories of human art are simply pieces of biological stuff fizzing in my brain' was a fairly typical response (#73). The single exception was a science writer who told us, with some pride, that 'I personally try to live without believing in free will, and it's taken me most of my life' (#24). Beyond that, and irrespective of whether they held religious beliefs or not, expert interviewees roundly attacked the view that these scientific disciplines proved that the basic building blocks of human existence were illusory.

Their main disciplinary target in all this was neuroscience, although the critique extended across the entire scientific front. Experience—all experience—had an undeniable neural substrate or material correlate in the brain. Every experience could, in theory, albeit with bewildering complexity,

218   THE LANDSCAPES OF SCIENCE AND RELIGION

be mapped onto the activity of billions of synapses. But that didn't mean anything about the reality of the experience itself.[52]

In a similar critique, evolutionary psychology might have 'damaged our understanding of humans, the human condition and human behaviour', as one biologist claimed, but if it had that was a problem for evolutionary psychology not the human condition (#35). Looked at through the 'lens of neuroscience' there was no evidence for free will. But, as the philosopher in this case went on to point out, 'what do you expect to see through the lens of neuroscience?' (#12). Similarly, consciousness was real, however much it failed to turn up on an MRI scan.

> We do have consciousness, we have self-consciousness, we can reflect on our own beliefs and desires. We have life plans. We feel guilt and shame and remorse and gratitude. We do all this stuff that no other animals do. Neuroscience can't possibly impugn any of that. (#64)

In a similar way, human moral sensitivity and reasoning may well have roots sunk deep in kin selection and reciprocal altruism, but that did not mean it was arbitrary or meaningless. Morality might be evolved, but that doesn't mean it's just 'a thing we made ... an artefact' (#64). One interviewee recalled confronting Richard Dawkins on this point, putting to him the challenge that 'your belief that rape is wrong is as arbitrary as the fact that we have evolved five fingers rather than six'.[53] In short, many of science's challenges to religious ideas in this part of the debate were challenges to basic human ideas, and to follow them all the way would be to abandon pretty much everything we believe about the human condition. Very few people to whom we spoke were prepared to do that.

There are three possible explanations for this reluctance. First, myopia. These scientists, philosophers, journalists, and theologians, like almost everyone else, were simply trapped in their own worldviews, unable to see the light thrown on the human condition by science. Second, denial. The intellectuals in question *were* able to see the light thrown, they were simply unable to deal with it, unwilling to face the unnerving reality of human life revealed to them by evolution and neuroscience, clinging to outdated notions of the self and consciousness and moral reflection that were no more credible today than ideas of the soul.

Either of those is possible and some well-known authors, such as Yuval Noah Harari, have made a name for themselves in positing a combination of the two. There is, however, a third possible option on the table, one that has a significant bearing on the wider science and religion debate. Perhaps science is simply the wrong tool to identify the kind of thing we have been talking about in this chapter.

Expanding on what the philosopher quoted earlier said, if neuroscience (or evolutionary psychology, or anthropology, etc.) cannot discern any of the things that have long been and are commonly accepted as basic to human nature—that we are coherent selves, that we are conscious, that our experiences are real and not simply in the mind, that we have genuine agency and some form of free will, that moral reflection is not simply an illusion, that ethical judgements have meaning and content rather than being arbitrary and entirely contingent on evolutionary chance, etc.—if science cannot discern any of these things, perhaps it is science that is at fault? Or, less provocatively, perhaps the scientific disciplines in question are just the wrong vehicles for conducting such conversations? Perhaps the perceived tension in this part of the science and religion landscapes, is actually a bit of a storm in a teacup, borne of two 'disciplines' that are ultimately talking different and incommensurable languages?

Outstanding questions

This is a tempting place at which to conclude this particular discussion. As we intimated at the start of the chapter, in spite of the popular attention that this part of the interaction between science and religion gets—all the headlines about God or the Big Bang, evolution or creation, neuroscience or spiritual experiences—we believe that (1) so many of these debates over 'content' pivot ultimately on different perceived anthropologies; the content is really the content of our understanding of the human; and (2) this part of the 'conflict' is, in fact, less abrasive and more tractable than some other areas of tension. That was certainly the view to which the expert opinions tilted. Perhaps because they wanted to distance themselves from popular notions of the science and religion conflict, even those interviewees who tilted towards some fundamental methodological or metaphysical tension between science and religion tended to downplay the conflict here.

However, to end here would be dishonest, not simply because there were people to whom we spoke who located conflict here but, more pertinently, because some of their reasons for doing so do strike us as profound challenges, not amenable to apologetic responses, and to

220   THE LANDSCAPES OF SCIENCE AND RELIGION

which religious thinkers should (continue to) give serious consideration. Put another way, the alleged conflict between science and religion
here may indeed turn out to be largely smoke, but there are still some
well-aimed shots being fired within this fog of war. We would like to
highlight two, both of which have been mentioned in passing in this
chapter.

The first is the challenge of gradualism. There are no clean breaks
in evolutionary history, at least none that we can see. There is no single point at which protohumans became human, or became conscious,
or morally aware, or cognisant of their mortality. By implication, there
is no firm boundary you can throw around the idea of 'the human'.
Indeed, one might make an argument for this category of the human
being as fuzzy a category as science and religion, with none of its
potential disambiguated elements—relationality, morality, agency, consciousness, reason, etc.—being necessary or sufficient to determine the
category.

And yet, most religions, and certainly those familiar in the West, do have
a notion of the human, a category that, for whatever reasons, merits a different and privileged kind of attention. For Abrahamic religions, there is a
category of creatures who have special responsibilities, who are capable of
a particular relationship with their creator, and who can possibly look forward to a post-mortem existence. None of this is to deny their thoroughgoing
interest in the rest of the creation. It is simply to recognise that 'the human'
is not only a meaningful but also a necessary category for such religious
systems.

These two different approaches do rub up against one another uneasily,
which is why so many believers and theologians have resorted to the idea of
ensoulment to demarcate a difference. That is, however, a problematic idea,
certainly when it comes to a scientific understanding of human history (or,
for that matter, human conception).

> There was no sudden point where you can say "Ah, that's it, we've become
> human, we've become conscious of the divine." This is what some religious
> people have said to me, that they can accept there's been evolution, but at
> some point, God put a soul into humans and not into any other animals.
> Only humans have souls and that must be an intervention by a divine force.
> I can't see the evidence for that, so for me, it's a natural process without any
> evidence of a steering hand or intervention. (#72)

In truth, this problem is simply a subset of the wider problem that attends all taxonomies. Nor is it fundamentally different to the long-standing debates about when exactly we can say a human life begins: conception, implantation, sentience, viability, etc. The question of when and how we trace the origins and definition of humanity is not dissimilar to the question of when and how we trace the origins of individual humans.

Put in concrete terms, this whole area does pose certain questions of those religions that have a particular recognition of humans, questions like: when did that recognition begin in human (pre-)history? Does it extend to the billions of fertilised but non-implanted embryos that have 'existed' invisibly throughout human history? Does it extend to those non-human animals that exhibit sociality, intelligence, reason, and perhaps even morality and agency? Would it extend to the potentially billions of other life forms strewn throughout the cosmos that show similar signs of consciousness, agency, and morality? None of these is, we think, an insuperable problem for such religious systems, but they do lay serious questions, that come from a broadly scientific provenance, at the door of those systems.

The second and surely the more significant challenge here comes from the understanding of 'creation' that is revealed by the theory of evolution by natural selection. This is a well-trodden territory, and it is important to be clear about the exact nature of the challenge here. This was not that evolution by natural selection had revealed a creation a great messier and more 'spatchcock' (#61) than had heretofore been thought. That was a problem only really to those religious thinkers who still adhered to a kind of Paleyian natural theology, in which natural was deemed flawless in a 'design and engineering sense' of the word.

Nor was it that natural selection was 'wasteful'. This was a common accusation but one that carried with it certain assumptions that were often questioned by interviewees. The 'idea of wastefulness' was a problem only to 'the part of us that worships utility and efficiency' (#75). What to one perspective is waste, to another is 'abundance' (#84). Moreover, given that pretty much everything in creation was re-used, time and time again, to call anything truly wasted seemed to stretch the meaning of that term.[54]

Nor was it even the existence of pain. Plenty of interviewees pointed out that pain was essentially a response to negative stimuli without which no sustained life was possible. 'Where we see people who don't feel pain, who don't feel suffering, either they don't feel social suffering and they become

222 THE LANDSCAPES OF SCIENCE AND RELIGION

psychopaths, or they don't feel physical suffering and in fact it's a curse and not a blessing' (#85).

Rather, it was that evolution revealed (1) a scale and (2) a necessity of pain and suffering in the world that was hard to reconcile with any understanding of a benevolent (still less a benevolent and omnipotent) God. The first point was made powerfully by one philosopher:

> The depths of pain and suffering that you see is so appalling and then when you realise that's going on across the entire face of the planet and has been going on for millions of years, we're looking at pain and suffering on such a vast scale, revealed by science, that for many of us, it's just not plausible that this is the creation of a supremely powerful and benevolent creator who has a special place for us in this universe. (#42)

The second was made by a theologian:

> If God is good and is the creator of this world, why create it with so much death and suffering, not as accidental pieces of the creation puzzle but as actually the drivers of evolutionary development, the very skills and beauty that we so admire in nature? (#85)

To be sure, answers to these were offered. Beauty came from the pain. 'Altruism' resulted from the 'competition'. God suffered with his creation. Perhaps this was 'the only way that God could have given rise to this sort of world' (#100). Such explanations are more or arguably less effective. But conspicuous by its absence from among them was the traditional Christian answer of 'the Fall'.[55] No one among our interviewees posed the existence of a pre-Lapsarian existence or an actual (pre)historical Fall as a means of explaining the pain and suffering, presumably because science no longer left that option open.[56]

The result was, at the very least, a difficult challenge laid at the door of (in particular) Christian doctrine. The world was not, in original intention, a pain-free, suffering-free, death-free, pre-Lapsarian thing but one in which pain, suffering, and death, on a vast scale, were built into the whole process, and this was a fact that was hard to square with a good God who made a good creation. In the words of one theologian:

> Christian theology needs to be much more honest about the extent to which the God who is the creator of everything else that exists whatsoever

is therefore responsible for this world, which is not only a suffering-filled world but a world in which suffering in a measure drives the developments of creaturely characteristics. (#100)

Notes

1. See *Exploring the Spectrum* and Unsworth, Amy and Voas, David, 'Attitudes to evolution among Christians, Muslims and the non-religious in Britain: differential effects of religious and educational factors', *Public Understanding of Science* (2018), 27(1), 76–93.
2. Miller, J.D., et al., 'Public acceptance of evolution in the United States, 1985–2020', *Public Understanding of Science* (2022), 31(2), 223–238.
3. Barnes, M.E., et al., '"Accepting evolution means you can't believe in God": atheistic perceptions of evolution among college biology students', *CBE: Life Sciences Education* (2020), 19(2).
4. His other pinch points—about which institutions ('religious' or 'scientific') get to set the moral purpose and meaning of a society, and how medical technology, such as that dealing with embryonic stem cells or genetic modification, should be used and regulated—correspond closely to our fourth area of disagreement, which we will explore in the next chapter.
5. The other is the way in which Americans have serious questions about what 'science' means for the existence and activity of God—a metaphysical issue we touched on in an earlier chapter.
6. 'The first problem with cosmology is that it contradicts the kind of very basic creation myths of most religions. But as I said before, it doesn't need to be a point of conflict if you don't take those stories literally.' (#15)
7. 'The Big Bang actually makes the atheist cosmologists twitch. You want a moment of creation? I can't think of a bigger one and so that's really a bit of a freaky thing for an atheist cosmologist to look at.' (#43)
8. 'You can actually create a universe from zero energy because the amount of pulling things apart creates negative energy in the form of gravity and that is equivalent in energy to the amount of energy you need to make the stuff, and so then you can get to the point of creating a universe out of quantum fluctuations.' (#1)
9. 'He doesn't want to invoke God and no physicist wants to invoke God really and the multiverse is a way to fine tune the universe without invoking God.' (#32)
10. 'Terence McKenna called this view the view where scientists say give us one free miracle and we'll explain the rest. And the one free miracle is the appearance of everything in the universe with all the laws that govern it, from nothing in a single instant.' (#41)
11. For example, from one eminent cosmologist who said of multiverses, 'some people say that the multiverse idea is not part of science because we can't directly observe these other universes ... but I think that's an argument which I think has weaknesses. If you have strong reasons to believe a theory, supposing this unified theory predicted the existence of gravity, neutrinos, and everything, and got that all right, they regain credibility and we would then take seriously what it predicts about context where we can't make direct observations.' (#33)
12. In a similar vein, 'I don't find any cosmological argument threatening, the multi-verse, cool, that's fine. My theology can hold any of those options.' (#23)
13. In a similar vein, 'A lot of cosmology is theoretical, just as theoretical as religious explanations in some ways' (#29). This is 'an area where the science does still struggle to provide the whole story and where it seems that in the end, the science itself is confronted with philosophical questions' (#10).
14. See Spencer, *Magisteria*, chapter 6, for more on this.

224   THE LANDSCAPES OF SCIENCE AND RELIGION

15. 'Humanity will not exist on Earth in a million years, but there will be new type of lifeform on Earth.' (#27)
16. 'It has to be said, you're talking about extreme time scales. In principle you can extract energy from black holes, so it's whether you care about the 10 to the 120 or not. We're into a strange kind of debate that to me doesn't mean anything, but I see the principle is important.' (#77)
17. 'If we discovered it there [on Mars], we would know immediately that life was commonplace throughout the entire cosmos because of the extraordinary fact of it occurring twice in such close proximity.' (#54)
18. For example, it could be interpreted as showing that the universe was made for life, intelligence, and possibly even love, thereby implying that there was a loving intelligence or mind behind it.
19. When interviewed in 1999, Stephen Jay Gould agreed that creationism was simply an American phenomenon. See Dreifus, Claudia, 'A conversation with Stephen Jay Gould: primordial beasts, creationists and the mighty Yankees', *New York Times* (21 December 1999).
20. 'You see bits and pieces in the newspapers about faith schools, but I don't have any direct experience of that. I was a school governor, and it never came up at our school.' (#29)
21. 'I think for biologists and particularly young evolutionary biologists who want to break into the media, it's a useful and sort of soft springboard and I think a lot of the pattern is often the same as the one I did, which is there is a soft target here which I can write about passionately and it's very popular.' (#35)
22. 'It is all tied up with politics and right, left and culture war stuff. It's bizarre to me that you can predict what someone will say about evolution for example, by finding out what they think about ideal tax policy.' (#40)
23. To be clear, while 28% of people disagreed or strongly disagreed that 'the theory of evolution is unable to explain human consciousness', only 27% of people agreed or strongly agreed with that statement. A further 26% neither agreed nor disagreed. In other words, public opinion was divided (and confused) on this issue rather than anti-evolutionary in any coherent way.
24. Again to be clear, while 27% of people disagreed or strongly disagreed that 'the theory of evolution cannot explain the existence of morality', only 28% of people agreed or strongly agreed with that statement, and 27% neither agreed nor disagreed. As with consciousness, public opinion on evolution and morality was primarily divided and confused.
25. Although this was a far bigger tension at the time of evolution's emergence. 'The reliance in many religious circles on the design argument ... was the real problem in the nineteenth century, that evolution undercut people's faith that they could prove God through the design argument. Now if you actually are trying to prove God through the design argument evolution just undercuts that completely.' (#16)
26. Even here though, the example given—'the way evolutionary anthropology looks at the archaeological record, it does not leave space for big unanswered questions that religion would address' (#78)—gestures towards the anthropological heart of this matter.
27. 'It is hard to see why a loving God would ... use such a long-winded gruesome process to create.' (#17)
28. 'When you start to get into the logic of theistic evolution, it very quickly departs from what most scientists are talking about when they're talking about evolutionary processes.' (#23)
29. 'There seems to be no guiding hand there. It's all running as a natural process from that first beginning. At no point is there any apparent divine intervention.' (#72)
30. 'Catastrophic events overcome vast numbers of them, 90% of maybe species disappeared at some of these catastrophic intervals and then life carried on in a whole new direction until we arrive at our present world.' (#72)
31. 'If you were the creator that designed humans, you really didn't do it very well. It's just not very good, it's just a mess.' (#77)
32. 'If your religion tells you that humans are separate to other animals that is going to be a block.' (#9)

ANTHROPOLOGY   225

33. 'I think one of the big problems is that humans consider themselves to be special.' (#15)
34. In a similar vein, 'it's very hard to see from a physics point of view how there can be free.' (#9)
35. 'What we now think of as religion, would have been our protoscience and it would have been accompanied by a proto-technology, which is propitiation of these forces, sacrificing to them or pleading with them or praying to them to stop the rain or bring the rain, to help you defeat your enemies and what have you.' (#61)
36. 'There's lightning in the sky, someone's died. Why have these things happened?' (#72)
37. 'It may be the belief system started as a way of countering anxiety. Worry about what happens after death. Where do we go, what happens? The reality of facing death. It may have been a way of, if you like, a comfort.' (#72)
38. 'I think religious people take offence at the idea that their beliefs are simply a by-product of the way the evolved mind works [but] ... I think that's a really good explanation for religious beliefs and for religious phenomena.' (#13)
39. A neuroscientist who studies the relationship between brain function and various mental states, and who has made a particular focus of the neurological study of religious and spiritual experiences.
40. Others deployed the classic argument against the experiment's circularity. 'There are a lot of methodological reasons why that is wrong ... what about Dr. Libet? Is he free in this experiment?' (#12) On why the experiment was flawed, see Taylor, Steve, 'How a flawed experiment "proved" that free will doesn't exist', *Scientific American* blog (6 December 2019.)
41. The interviewee continued: 'This approach to evolutionary biology could, at least in my view, be compatible with theo-dramatic understandings of how God acts in the world in relation to human beings and other creatures. Because it gives some sense of agency, and the importance of agency to particular beings' (#82).
42. 'The most hardcore molecular biological realm of biology, which deals with the *Caenorhabditis elegans*, a small nematode worm, it's now completely well-established that if you expose them to food that is flavoured with benzaldehyde, a compound they'd never normally encounter, for four generations, so they learn that they wriggle towards benzaldehyde because that's where food is. It's like a conditioned reflex. For at least 30 generations thereafter these worms will wriggle towards benzaldehyde, not because they've had a mutation in their gene that makes them sensitive to benzaldehyde that's selected over hundreds of generations, but because they've inherited this preference for food with benzaldehyde through epigenetic inheritance. No one knows quite how it works [but] ... everyone agrees that epigenetic inheritance is happening. Whereas until the year 2000, almost everyone in the scientific world denied, got terribly angry at the idea that it's impossible.' (#41)
43. 'I am interested in the evolution of elephants on islands, and because when elephants are on islands, they always become smaller over time. So, it is a phenomenon of repeated and parallel evolution towards this small size, dwarf elephants.' (#92)
44. 'We have two puzzling instances of bodies being deposited in caves. We have Atapuerca in Spain, that is probably around 400,000 years ago. There, as many as 30 or 40 individuals were put down a shaft into a particular cave. There are no signs of occupation in the cave. It would be very difficult to get in and out. There is no carnivore activity. It is a very puzzling thing, and it seems to be that the bodies have been put in there deliberately.' (#66)
45. 'We are an unbelievably imaginative species and Robin Dunbar would explain it as our levels of intentionality and so on but we do create these imaginary worlds and then live in them and if you pin me down as to what the hallmark of humanity might be, it would be that and we're constantly looking at those ways to create and refresh those imaginary worlds and we do it through material artefacts.' (#52)
46. 'We've crossed the threshold where we start to see the meaning of things rather than just taking the world literally at face value every moment of the day, [and this] gives us the capacity to contemplate things that we can't see.' (#45)

226 THE LANDSCAPES OF SCIENCE AND RELIGION

47. Or, in the words of one Christian theologian, 'I've got no problem with humans not being unique in the way that we've previously thought … I think we're special because God has chosen us and we know that in Christianity because God decided to become a human being rather than a tortoise' (#83).

48. It seems that no one felt confident enough to have the same conversation about the vis-à-vis other religious traditions.

49. Polkinghorne, John, 'Eschatological Credibility: Emergent and Teleological Processes', in Peters, Ted, Russell, Robert John, and Welker, Michael (eds.), *Resurrection: Theological and Scientific Assessments* (Grand Rapids, MI, 2002), p. 51.

50. 'It's very hard to see *from a physics point of view* how there can be free will because if it's there, surely it has to be there at the level of fundamental forces' (#9); emphases are added because the phrase is so significant.

51. 'I'm not denying that there is consciousness, I'm just saying that you don't need to suppose that there's anything non-physical going on to understand.' (#55)

52. 'It is a category mistake to suppose that anything that you can learn about the brain can one way or another influence what you believe about the conscious self. All we know is correlates, we know there are correlates that often are very closely connected and when you alter one, you will alter the other. That's all we know in this world.' (#75)

53. The full exchange, as remembered by this interviewee, is worth reading: 'I said to him, "Look, if we had evolved into a society where rape was considered fine, would that mean that rape is fine?" And he said, "I don't want to answer that question. It is enough for me to say that we live in a society where it is not considered fine." … And I said, "But when you make a value judgement, don't you yourself immediately step outside of this evolutionary process and say, 'The reason this is good is that it is good', and you don't have any way to stand on that statement?" And he said, "Well, my value judgement itself could come from my evolutionary past." And I said, "Therefore, it is just as random as any product of evolution." And he said, "You could say that" … And I said, "Okay, but ultimately, your belief that rape is wrong is as arbitrary as the fact that we have evolved five fingers rather than six?" And he said, "You could say that, yes." Which I thought was a real interesting admission' (#44).

54. 'I don't think that looking at it through an ecological lens you can actually say it's wasteful at all. Every death springs life to numerous other creatures so yes, it's a difficult kind of argument to make from an ecological perspective, that nature is wasteful.' (#85)

55. One interviewee, with a particular expertise in this area, was clear that 'if you look at the tradition of interpretation, up until about the Reformation, people didn't think that all death entered through the Fall.' (#85)

56. We are conscious that there are some, often eminent, Christian thinkers who *do* hold to this view or even that of a cosmic fall, laying all the suffering of creation at the door of angelic rebels, but neither of these views were mentioned by our interviewees except occasionally to dismiss them.

10
Public Authority and Public Reasoning

Introduction

According to the ruling in *Kitzmiller v. Dover Area School District* in 2005, science 'does not consider issues of "meaning" and "purpose" in the world'. Instead, it diligently overlooks 'ultimate explanations' in pursuit of what can be inferred from 'empirical, observable and ultimately testable data'.

The judge in the trial came close to stating what the other legal and institutional definitions we explored in part I implied by omission: science is not to be defined by its orientation towards any particular social or ethical stances towards the world. Its very openness—its tentativeness (Overton), its commitment to critical analysis (Science Council), its universalism (Merton), its epistemic humility (disambiguated 'family' definition)—should preclude it from being owned by, or from lending unreserved support to, any group, ideology, position, policy, or programme.

In theory, the same applies to religion. None of the family features in the disambiguated definition, or the formal definitions given in US or English law, or the various academic understandings we looked at in part I commit religion to a *specific* social or ethical stance. They do not, of course, preclude such stances. The mere fact of holding ultimate concerns (Tillich, *Malnak v. Yogi*), or possessing an ethical or legal dimension (Smart), or impacting the community (Charity Commission Analysis), or having a formalised institutional presence (disambiguated 'family' definition) means that it is almost certain that religion will have some social or ethical identity, likely a potent one. But which identity, and what that means in practice, is not prescribed.

That being so, there need be no conflict in this part of the science and religion landscape. Science isn't an ethical or social position, and religion has no prescribed ethical or social position. The two should not, or at least not necessarily, be in conflict with one another here.

If this is true—and, at the risk of spoiling the plot, that is the conclusion to which this chapter gravitates—this does not mean that people do not think that science and religion are in conflict here. Indeed, a growing body of research is revealing how positions within the science and religion debate do

228   THE LANDSCAPES OF SCIENCE AND RELIGION

correlate with conflicting ethical, social, and political positions and agendas. In 2016, Shiri Noy and Timothy O'Brien explored how perspectives on science and religion mapped onto public attitudes concerning a wide range of social, political, and economic issues in the US.[1] There was already work showing that opinions on controversial topics of public interest, such as stem cell transplant, varied according to people's views of science and religion. Noy and O'Brien showed that, in addition to these familiar issues, there are notable difference of opinion between what they called 'moderns', 'traditionals', and 'postseculars'[2] on a whole range of other issues, differences that were there even when other political, denominational, and class differences were accounted for.[3]

Noy and O'Brien's 2016 paper analysed data from five waves of the General Social Survey (GSS), covering 2006, 2008, 2010, 2012, and 2014. Four years later, they extended their analysis to 30 waves, ranging from 1973 to 2018, on the basis that, over this near half-century, science and religion had been enlisted in a growing number of issues, such as bioethics, sexuality, race, and environment, often as moral and political alternatives to one another, and that the process had 'gradually transformed their cultural meanings'.[4] They found that politics had grown 'increasingly entangled with how Americans think about science and religion', with confidence in science displacing confidence in religion within the Democratic Party and vice versa for the Republican Party.

This approach not only underlines the idea we have pursued in this book, that science and religion are culturally mediated entities and need to be analysed through the lens of how the words/categories are actually used in society, but also further stresses how *politicised* that mediation is. However much it might be the case *in theory* that science isn't an ethical or social position and religion has no defined ethical or social position, in reality—in the US at least—perceptions of both have become 'increasingly rooted in moral and political dispositions', are increasingly 'politically divided', and this has 'contributed to a belief that they provide not just alternative sources of cultural authority but opposing ones'. In effect, and in line with what John Evans has argued, Noy and O'Brien believe that not only is there a battle underway in this part of the science and religion landscape, but also it is one of the fiercest we have seen. When people disagree about science and religion, they are disagreeing about the (perceived) profound ethical, social, and political ramifications of each.

The UK is not the US. Not only has there been less research on the political dimension of science and religion in Britain, but the culture is (so far!) less politically polarised. Nonetheless, we found the seeds of this division within our research. In among the different associations with the word revealed in PAS 2019, one in ten people voluntarily connected 'science' with 'advancement/progress/the future/better world/helping mankind/easier living', in other words with a broadly progressive social narrative.

By contrast, in our quantitative research, while one in twenty people associated religion with the positive-sounding 'support/community', the same number connected it with the more conservative category of 'culture/tradition/civilization', and more than one in five did so with either 'war' (19%), 'hatred' (2%), or 'cause of problems' (1%). As we observed in chapter 6, largely without personal or educational experience of science or religion, certainly since school days, UK public perceptions of both are heavily informed by the media-narratives in which they are embedded, and while those narratives are not as partisan in the UK as they have become in the US, there are still traces of their impact on the public understanding of science and of religion.

These findings are supported by recent research conducted for the University of Birmingham's Exploring the Spectrum project. This sampled public opinion in seven countries (Argentina, Australia, Canada, Germany, Spain, the UK, the USA) and found that in all but one, it is more common to view religion as being, on balance, negative for society (the exception was America). In the UK, for example, 50% agreed that 'religion often has more negative consequences for society than positive consequences' and 30% disagreed, whereas only 13% agreed that 'science often has more negative consequences for society than positive consequences' while 63% disagreed. In a similar vein, in every country surveyed, a majority of people thought scientists were a reliable source of information (81% did in the UK, only 5% said unreliable), whereas in every country, this time including the US, only a minority thought religious leaders were a reliable source of information (27% in the UK said reliable compared to 42% who said unreliable, the rest saying 'neither' or 'don't know').[5]

To pick up a theme that will run throughout this chapter, there is no necessary conflict in all this—just differences of opinion and esteem—but it is not hard to see how these perceptions seed some kind of tension. When the UK general public was asked its view on various contentious technologies and scientific advances, such as stem cell research, genetically modified (GM) crops, and nanotechnology, those that we might call 'science publics' (meaning those who had a higher terminal level of science education, or had

230 THE LANDSCAPES OF SCIENCE AND RELIGION

a higher confidence in their level of science knowledge, or scored higher on the science knowledge quiz) were more likely to say the benefits outweighed the risks than were 'religious publics' (meaning those who claimed to belong to a religion, attended religious services, or held a conservative view of a sacred text). Those inclined to align their identity with science found themselves in implicit tension with those inclined to align their identity with religion. One sociologist of religion captured the nature of the encounter perceptively.

> In practice, in real life, what we see are arguments where there are religious believers on one side and secular scientists on the other, and they are disagreeing about the right way to go, whether it's to do with in vitro fertilisation, or testing of embryos, or things happening at the end of life, or whatever. And that these are real and thorny problems where you're seeing some kind of *de facto* divide. But I think it is *de facto* as opposed to in principle. (#28)

It was a view echoed by several other sociologists.[6] Whether or not people *should* be disagreeing between science and religion in this area, they were.

This chapter looks in greater detail at this area of apparent conflict. It begins by setting out the range of issues on which there was perceived to be conflict between science and religion. Interviewees, or at least the majority of them, were clear that conflicts here were neither universal nor necessary, but even having entered those caveats, it was clear that many sense some conflict between science and religion in these areas.

The chapter then moves on to explore what precisely underlies that perception of conflict, what deep ethical or existential orientations did science and religion have which so often seemed to generate these public tensions. It proceeds to analyse the nature and legitimacy of this conflict, explaining why this underlying tension was *de facto* rather than a necessity. Having argued this, however, in the fashion of the last three chapters, the chapter concludes by moving on to where the deepest fault line in this area lies, which, we argue, has less to do with issues themselves, or even the respective ethical orientations that people sense within science and religion, but are embedded in the fundamental question of public authority—who has authority to adjudicate on issues of shared public concern—and the attendant question of public reasoning—on what basis.

Battles over abortion, and euthanasia, and genetic engineering, and …

At the time of our research, married women in the UK had been able to have the contraceptive pill on the NHS for nearly 60 years, and non-married women for six years fewer. Medical abortion had been legal in the UK for just over 50 years, and in vitro fertilization (IVF) successfully practised for 40. The first human embryonic stem cells had been grown in a laboratory a little over 20 years earlier. Mitochondrial replacement therapy, in which mitochondria of a future baby are replaced by mitochondrial DNA from a third party, thereby giving rise to so-called three-parent babies, had been legalised five years earlier. Jennifer Doudna and Emmanuelle Charpentier won the Nobel Prize for Chemistry for their development of CRISPR gene editing during our research period. And physician-assisted suicide remained illegal. Each of these procedures had been enabled by modern science. Each was understood as a significant step forward in human well-being. And each was perceived, by some at least, as having been—and still being—fundamentally opposed by religious groups, institutions, and beliefs. Herein lay the conflict.

Abortion was perhaps the most obvious. Opposition to this was incomprehensible, literally so for some interviewees. 'I really believe in euthanasia and abortion as things that should be accessible to all as a human right. I think that if we were just plonked on the planet tomorrow as blank slates who were allowed to make up these new rules, everyone would agree with that' (#38). Ecclesiastical opposition was all the more incomprehensible in the face of what science revealed about the sheer number of fertilised eggs that did not go on to become completed pregnancies. 'Why isn't the Catholic church caring about the fact that the body naturally aborts one in two, two in three conceptions? Shouldn't it be giving you drugs so that you can save those lives, even if they are hideously malformed?' (#39). The Catholic Church's stance on artificial contraception was, if anything, even less intelligible. 'I have never really been able to understand [it] … the Catholic Church is prepared to produce an app which makes it easy for you to have sex without getting pregnant but continues to say that condoms are bad' (#21).

IVF was thought to have survived and flourished in spite of early religious opposition. 'We've got 10 million people on the planet who wouldn't have otherwise existed, for the amount of happiness caused by IVF … [but] it could've been banned by the fundamentalists back then'

(#50). That said, religious opposition to IVF was far from over,[7] and there was a new front opening up in this battle, in the form of stem cell research. '[There are] particular doctrinal objections to that, such as in the Catholic church, for example, [concerning] the dignity of early embryos, even though for scientists, that early embryo is just a couple of cells' (#82).

The scientific understanding of human sexuality had come on a long way in the 50 years since the American Psychiatric Association removed the diagnosis of homosexuality from the second edition of its Diagnostic and Statistical Manual (DSM), but the distance travelled by religious belief here was (felt to be) rather less. 'A lot of people still think that most religions still have a very conservative sexual morality ... but if you take a scientific world view to be one in which one takes a starting point from evidence and facts and all that kind of stuff, people see no reason why therefore certain types of sexuality should be deemed immoral' (#3). Antiretroviral treatment for HIV AIDS had been effective for around a quarter of a century, but at least according to one sociologist who had conducted research among them, 'I have heard many a conservative American fundamentalist saying openly that they regret that we have managed to cure HIV' (#21).

The issue that most vexed interviewees, however, was euthanasia, enabled as a safe, reliable, and dignified practice by modern science and banned, even in the face of public approval, by religious groups. According to a number of interviewees, reasons against physician-assisted suicide—'because the law could be abused or something'—were disingenuous. 'The reason they are against it is because of a Christian tradition which holds the soul to be sacred, and it holds the notion of suicide to be deeply disturbing and problematic' (#22)—a reason that had no scientific credentials and was in profound tension with the method and metaphysics of science. Here then, was the source of the science–religion conflict in this area of the landscape: science was capable of achieving a great many things for human life and dignity, and religious belief and groups repeatedly and quite unjustifiably opposed it.

To be fair, a number of interviewees pointed out that the traffic was not all in one direction here. Sometimes, though admittedly less today, there had been tension between science and religion for which science, rather than religion, was 'at fault'. Interviewees cited eugenics most commonly as a historic front in this particular battle.

There's a lot of emptying of skeletons out of cupboards at the moment, going on in science, in terms of seeing the tendrils of eugenics through things. And it's [also] worth pointing out that the one organisation that has consistently had clean hands in that respect ... is the Roman Catholic church. (#96)

Others mentioned experiments conducted on African Americans,[8] and the Nazis naturally came up. When science and religion found themselves squaring off here, it was not always religion that was morally culpable.

Whoever was to blame, though, there was a common impression that science and religion were in some form of conflict over these matters. It was an impression supported by our quantitative research, which asked people the extent to which they thought various scientific procedures were morally acceptable.[9] Time and again, 'scientific publics' were inclined to find such procedures more acceptable than 'religious publics'. To take one example, when asked about 'using and then destroying human embryos if doing so helps scientists find cures for disease', 36% of self-affiliating Christians, 23% of those who attended a religious service more than once a fortnight, and 16% of those who held a literalist view of the Bible thought the process was morally acceptable. This compared with 55% of those who had a terminal science education of degree level or above, 57% of those who felt confident in their knowledge of science, and 61% of those who scored highly on the science knowledge quiz. There was a similar difference identifiable in more speculative scientific developments, with 'scientific publics' being more open to new technologies, such as self-driving cars, artificial intelligence (AI) friends, and scientific immortality, than were 'religious publics'.[10]

The differences were not always pronounced. Only a small minority in both religious and science publics thought that 'creating a baby that is a clone of another person' was morally acceptable, and only a small minority in both groups thought 'using genetic engineering to create a baby that is smarter, physically stronger, or better looking' was OK. Differences in opinion between the two groups did not override the generic social antipathy when it came to contentious issues of this nature.

Nevertheless, the pattern discernible here supports many of the comments of the expert interviews, and is in line with the wider research alluded to earlier in this chapter, namely that there is an ethical–social front in the science and religion landscape. As we have already indicated, at least as many

234 THE LANDSCAPES OF SCIENCE AND RELIGION

experts to whom we spoke questioned the nature, solidity, and above all the necessity of this front. However, before we proceed to these corrections, we need to dip beneath the surface of these disagreements to see why science and religion are perceived to be in conflict here.

The difference over 'givenness'

Science isn't an ethical–social position, and religion has no prescribed ethical–social position. In theory the two need not be in opposition. Yet, as we have seen, some form of conflict is perceived here, running primarily, if not exclusively, through the rocky terrain of beginning and end of life issues. Why is that? Why could one theologian feel justified in saying, 'the ethical and the political ... just about 90 percent of the controversy is about that' (#83)?

The reason lies in the different 'stances towards the world'—the most generic phrase we can think of—that are perceived to be inherent within each of science and religion. In effect, science was (often thought to be) characterised by a certain attitude to the manipulation and control of the material world, and religion by a very different one. These views could be expressed through subtly different dichotomies. The most common was 'progressive v. conservative'. Within religious communities, we were repeatedly told, there is a 'tendency toward ethical conservatism' (#5). The point was not always made critically. Religion has 'a voice of slowing things ... [it] forms a useful societal function: just because something can be done, doesn't necessarily mean it should be done' (#62). Even those scientists who disliked 'reactionary voices ... holding them back' could still see them 'as being a valuable contribution to the ethics of society' (#13). However welcome religion's perceived conservatism was though, science was widely understood as being intrinsically progressive, if only in as much as its work advanced human knowledge and also therefore our ability to control and manipulate nature.

'Progressive v. conservative' easily slipped into 'modern v. archaic'. Religious dietary laws were a good example of this. The Jewish and Muslim proscription against eating pork could be 'understood in circumstances, in climates, in conditions where pigs are riddled with parasites'. Presumably, the philosopher in question mused, the original edict may have been a 'health proscription', but religion's conservatism 'transmuted' that into 'a blanket prejudice against pork' for which 'the logic ... no longer exists' (#61). In this way, the proto-scientific understanding that supposedly characterised religion was superseded by science's better understanding of nature, which meant that religion's retention of its rules saw it decay into archaism.

Both 'progressive v. conservative' and 'modern v. archaic' framed the ethical–social tension temporally. Another way of understanding the apparent difference was between the different foci of the two. Religious 'morality [was] grounded in God and in the biblical text' (#65), and that meant that God was its ultimate focus. 'Dozens and dozens and dozens and dozens of prescriptions [in the Old Testament] have much much more to do with not stepping on the cracks, which really irritates the gods, than it has to do with looking after widows and orphans and keeping your promises' (#61). In contrast to this, although it was not claimed that science was necessarily focused on the human good, it didn't really need to be. Simply by dint of its metaphysical naturalism, science couldn't be grounded in, or therefore focus on, any divine concerns. By default, it was claimed, the two activities had fundamentally different approaches to living in the world—one about pleasing God, the other thoroughly indifferent to any notions of the divine.

A fourth framing saw the difference as between the 'static' or 'inflexible' nature of religion's stance and the more 'flexible' or 'dynamic' one that characterised science. Religious ethics, we were told, 'tend to be very concretised'. Religious communities believe that 'their ethic is the dominant one, the pure one, the perfect one, because it came from God' (#73). It was an approach that was 'very difficult to argue with' and admitted very little 'wriggle room' (#38). Science, by contrast, although it did not possess a comparable set of ethics, was grounded in a developmental understanding of reality. It advanced by means of the admission and consideration of different voices and ideas, rather than their exclusion or elimination. It was comfortable with a pluralistic approach to the world in a way that religion was not.

This framing sometimes tipped into a fifth, which stretched the opposition even further. The logic here ran that religion adhered to 'an absolute morality' (#18), an objective understanding of the moral world. In contrast, science showed the ethics 'emerge from the human brain' (#77), and this meant that they were necessarily an 'incredibly personal' and ultimately subjective thing (#18). Science demonstrated that there was no absolute, overarching meaning to reality, and 'if meaning is local and temporary then ethics are also local and temporary' (#77).

Not all the framing of the different 'stances' adopted by science and religion were damning of religion. One sociologist drew out a different

236 THE LANDSCAPES OF SCIENCE AND RELIGION

distinction that is worth mentioning not only because it was interesting, but also because it grounded the ethical difference between the two in the respective anthropologies discussed in the previous chapter. 'It's more fundamental than just an ethical tension. It's a view of what human health and wellbeing consists in', the sociologist in question reasoned. Back in the nineteenth century, in the UK and US, there was a 'much more holistic view ... of social health', one that was sponsored primarily by Christian organisations. This contended that humans were 'only healthy in relation to the conditions in which we live and in which we socialise', and so adopted a much more rounded view of human health. Various scientific, i.e. medical, advances later—antibiotics, surgery, anaesthetics, etc.—this was 'massively narrowed down'. The individual human person was now conceived 'in a rather mechanistic way', and health was 'purely about treating disease'. The consequences have been disastrous, with the big killers of today being things like obesity and (ironically) pharmaceuticals (especially in America). 'That kind of medicine is making us very sick' (#67).

This last dichotomy shows that not all of the perceived ethical–social differences between science and religion were treated as the cause and fault of religion. That said, as will be obvious from this preceding discussion, many were. In fairness, it has to be admitted that some of these critiques were not always hugely sophisticated, and that this topic elicited more and coarser caricatures than perhaps any other discussion we had. We were told how 'the church continues to chunter about homelessness and poverty and such things, and nobody pays the least bit of attention, but so much effort, certainly in western Christianity, is down to trying to control people's sex lives' (#21). We were told that religious people 'claim that if you are not religious, you cannot possibly have any morality because ... the only way anyone is ever moral is the carrot and stick approach that after they die' (#19). And we were told that 'religions will give you a black and white standard. "It's all right to turn the trolley to kill the apostates, but it's not all right to kill the good Muslims who've been doing their prayers five times a day and so on"' (#24). Such critiques are a salutary reminder that even education to a professorial level does not indemnify you against crude caricatures.

Such condemnations notwithstanding, there was a sense among the expert interviewees of a deep difference between scientific and religious stances towards the world, whether expressed as that between a conservative and a progressive approach, or a modern v. an archaic one, or static

PUBLIC AUTHORITY AND PUBLIC REASONING   237

v. dynamic, or even objective v. subjective. The difference was sometimes captured in the idea behind the phrase 'playing God'—science characterised by its ability to modify creation and religion by its reluctance to do so.

> Where I would have a problem is if somebody said, 'this shouldn't be done because it's messing with God's work or something like that ... the moment of creation of life is a God-given moment and we shouldn't do that.' Clearly the fact that we can do that I think negates that argument. (#8)

In fact, the crucial word in this was not so much 'God' as 'given', the idea that there was a givenness to life and nature which meant it couldn't be modified. This was the line that ran through artificial contraception,[11] abortion, genetic modification,[12] and euthanasia. '"Because my life was given to me by God and therefore only God has the right to take it and I don't". I find that a nonsense argument' (#46). This was where the alleged battles lines were drawn. But even as this line was drawn by some interviewees, it was being erased by others who insisted that, however much there might be a tension here, it was not a necessary or inevitable one.

Closing the gap

Both sides of this alleged local conflict—the idea that religion is inherently conservative when it comes to the way in which humans should interact with the material world, and the idea that science is unproblematically progressive in this area—are open to debate.

Plenty of interviewees, including those most inclined to condemn it, pointed out, rather basically, that religion is not necessarily conservative. 'Let's be fair to religion here', one commented, 'most major religions have produced liberal wings which do not much want to narrow people's behaviour, and are not in the business of being bossy and telling us what is pleasing to God and what isn't' (#21). One even pointed out that with regard to some of the most contentious issues of the moment, gender and trans rights, a number of religions were at the more progressive end of the spectrum.[13]

Moreover, it didn't really matter whether these more 'liberal' attitudes were deemed to be the authentic expression of the religion in question or simply a peripheral one. The mere fact of their existence meant that, *de facto*,

238 THE LANDSCAPES OF SCIENCE AND RELIGION

religions boasted different views on these issues, and this alone defused the necessity of any conflict here. Numerous experts, on both sides of the argument, made this point.

> It's entirely possible that one could take a different view and still be respectably religious ... there are a number of different understandings within the Christian church in some of these areas ... Religious people are not monolithic. They have various views. Some are in favour of abortion, some are against ... you'd find Christian ethicists on both sides of the debate, so it's not a religion versus science issue at all. (#28, #16, #42, #100)

Furthermore, as if this didn't complexify this side of the conflict enough, there was the fact that the actual belief and behaviour of self-confessedly religious people was often very different from their religions' official positions. Religious behaviour and formalised religious belief did not always match. Of course, for those for whom religion was characterised primarily by the *formalisation* of beliefs and the *institutionalisation* of commitment— the fourth and fifth features in our disambiguated family resemblance—this point is irrelevant, and merely an example of religious hypocrisy. In contrast, however, for those for whom religion was characterised primarily by *personal* beliefs and practices, it further defused the apparent tension between science and religion in this area.

Finally, but perhaps most significantly, was the nuance that surrounded the idea of 'givenness'. Recognising something—whether an individual life or the entirety of creation—as 'given' did not preclude its manipulation or modification. To be 'given' did not mean to be preserved from all interference, particularly if your tradition held, as the Judaeo-Christian one does, that (1) humans were made to be co-creators, acting as God's 'stewards' or 'vicegerents' on earth, and that (2) however 'given' creation is, it is also 'fallen', and crying out for repair and restoration.

Put another way, the instruction not to 'play God' was not cited by the theologians or religious thinkers we interviewed because they did not agree with it. On the contrary, if anything, the view among them was that their religious beliefs insisted that they *should* be 'playing God'. 'For me the Judaeo-Christian tradition says that science and technology are gifts that are to be used under the kingship of God, we are stewards of science and that we exercise science for compassion, health, healing and delight' (#16). In short, religion was, in neither principle nor practice, quite so well fitted to the reactionary and conservative role in this alleged conflict as some claimed.

PUBLIC AUTHORITY AND PUBLIC REASONING   239

The other side—the idea that science is progressive—was less open to debate. Whether or not it was tied to a particular socially progressive agenda, science did have a clear association with development and control and progress. But it was precisely this fact that meant it needed ethical restraint. Indeed, the more potentially progressive it was, the more restraint it needed. Science might indeed have been naturally progressive, but that didn't mean it was unproblematically progressive. Approached in this way, the potential opposition presented by religion to science in these social–ethical issues was not so much a matter of conflict but of necessary and productive tension.

However competent it was, science could neither inspire an ethical response nor dictate its nature.[14] Science, we were told, was too often 'retrospective in its ethics' (#29). The technological advances for which it is responsible repeatedly 'run ahead of the societal and ethical debate' (#29). It needed a more conservative perspective to bring it back into line.

There were other reasons to welcome alternative voices here. Scientific work was specialised, and scientists needed help from experts beyond their own discipline.[15] Science was, culturally speaking, ill-suited to the uncertainty inherent in ethical evaluation. 'There's a cultural problem in terms of our caricature of what science is, that limits how possible it is for a public scientist to speak with nuance because they're always seen as the voice of facts' (#81).[16] Science required 'ethical frameworks' that it could not generate itself (#40). It was 'descriptive rather than normative' and '[didn't] do normative questions very well, or indeed at all' (#1). It 'exsanguinate[d] the world in order to arrive at general principles', and while that made it 'powerful and able to make predictions', it did nothing when it came to establishing 'value or meaning' (#12). Ultimately, science 'tells us about means to ends and leaves open what the ends are' (#55).

To be clear, this did not mean science had nothing to contribute to ethical debate. 'Many ethical questions about medicine need medical data' (#11). Deciding whether blood donors should be paid, for example, was an ethical decision, but one that could and should be informed by scientific data.[17] More generally, if you had decided upon a utilitarian approach to an ethical issue, 'science is going to be incredibly useful to you' (#40).

However, science could not tell you whether 'it was better to protect ... the physical health of old people [or] the mental health of young people' during a pandemic (#96). It could not tell you how to value the existence, health, sentience, consciousness, or viability of an unborn child, or to how to weight any of those factors and the freedom and autonomy of a prospective mother when evaluating the right approach to abortion. It was of limited use

240　THE LANDSCAPES OF SCIENCE AND RELIGION

if you wanted to take a deontological or virtue ethical approach, as opposed to a consequentialist one, and it was next to no use in helping you to decide which approach you should adopt in the first place (#40).

> If we decide that we want to increase ...lifespan or if we want to try and increase happiness or whatever goal that we have in life, we want to act in a way that this outcome will happen, we can use science to work out what strategy would best serve that. But the ultimate kind of value that you choose initially of what kind of person you want to be, what kind of world you want to live in, what kind of effect you want to have, I don't think can come from physical measurements of things. (#36)

In short, we need an ethical counterweight to science because, as one philosopher said, 'progress is always a good thing if you are scientific ... [but] scientific progress may not be progress in the broader sense of the term' (#14).

This was where religion stepped in—although with a massive, neon-sign caveat firmly in place. First the place of religion, then the caveat. If the question of scientific progress naturally spilt over into that of social or ethical progress—'progress in the broader sense of the term'—and this spillage meant opening up questions of meaning and value to which science could not provide answers, then it was clear that there was a potential role for religion in this discussion.

In one regard, that potential lay not in their being religious per se but simply by not being scientific.[18] As one journalist put it, 'the last people I would want in charge of nuclear power are the nuclear power industry' (#29). External perspectives on contentious social and ethical issues were necessary and valuable, all the more so if (1) those perspectives were socially embedded, active in and representative of a civil society that was going to be affected by the decisions; (2) they were part of traditions that had demonstrated an effective (or at very least enduring) contribution to ethical debate over the years; and (3) they were 'sophisticated in their ethical and moral reasoning' (#10) and had, as one interviewee put it, 'a developed ethical vocabulary that helps people to think about ethical issues' (#5).[19] Particularly if these criteria were met, there was a strong case for a productive religious presence in these debates.

Furthermore, if science's naturally progressive inclination required a more conservative ethical and social counterweight, religion's potential role here could be all the more important, at least if the loose association of religion and ethical conservatism held. Thus, the familiar (though, in fact, as we saw, rarely voiced) objection to 'playing God' actually just articulated a natural, healthy, and widespread reservation concerning the potential futures generated by scientific progress.

> When we talk about ... the possibility, however remote ... of manipulating genes in order to produce more intelligent people or more competitive athletes, then we're getting into much more dangerous ground ... and my own response will be to say we should definitely not, because we don't know where it's going to lead. And I think that the religious person saying you shouldn't do it because that's 'playing God' is probably expressing the same moral feelings that [I am, but] providing, as religious people tend to for their moral beliefs, a God-based rationalisation. (#46)

Approached in this way, even those most antipathetic to the religious worldview acknowledged the potential value of the religious contribution to this debate.

Now the caveat. However much there might be a voice, even a valuable one, for religion in the wider ethical and social debates thrown up by scientific progress, that voice was one amongst many, shorn of the authority that was once, and sometimes is still, claimed by religious institutions. As a science journalist said, 'what scientists struggle with is that they think they [i.e. religious figures] have the moral authority to *decide* sometimes what is and isn't right' (#13).

In one respect, this should be so obvious as to be hardly worth saying. It was unclear who, beyond the arrogant, infallible theocrats of secular nightmares, honestly thinks the religious have the right to decide authoritatively and finally on complex ethical–social issues. That said, a number of interviewees did think it worthwhile to state that they didn't believe the religious should be able to 'impose' their views on them or on the wider public, although that word 'impose' was often made to do quite a lot of work, from describing a certified theocracy to condemning an opinion with which someone disagreed.[20] However much the theocratic nightmare might be fantasy, at least in twenty-first-century Britain it still holds power.

That nightmare position, however, obscures a more moderate, complex, nuanced, and realistic debate that remains live in this area, and one that, we believe, lies at the heart of the whole apparent ethical–social conflict between science and religion. It is, simply put, who has the right to decide on these complex issues? In the words of one theologian, 'Who has authority?

242 THE LANDSCAPES OF SCIENCE AND RELIGION

Who are the priests in a society? Is it people with neuro in front of their name or is it the Archbishop of Canterbury? Is it the chief medical officer ... or is it some religious professional?' (#84). Where does the public authority ultimately lie and, just as important, how is it justified?

Public authority and public reasoning

The historical relationship between science and religion, and particularly their moments of tension, has been heavily informed by the question of who—which disciplines, practices, and institutions—has greater claim to authority in social and political affairs.[21] Indeed, there is good reason to believe that the period of most pronounced conflict, at least in the UK, during the second half of the nineteenth century, was due in large measure to the waning of clerical intellectual and social authority in favour of the (newly baptised) 'scientist'.

In reality, even as that was happening, a new source of public authority was emerging in Western Europe. Religious authority, it was at the time claimed, was elitist, institutionalised, and alienated from the concerns of 'the people', but, in truth, the scientific authority that challenged it was hardly more 'of the people'. Members of the X-Club, the exclusive and influential scientists' dining club set up in the 1860s to wrest authority from clerical hands, were as antipathetic to amateurs (and to women), let alone to views of the general public, as they were to priestly authority. Not without reason did critics, then and now, accuse such scientific modernisers of replacing one clerisy with another.

Within 60 years or so, however, the UK had embraced full suffrage, and it was clear, in principle at least, where public authority lay: with the public. 'The only people who have any legitimacy to dictate to us what we can and can't do are democratically elected representatives ... scientists aren't elected, we don't choose them, so they can't have that level of authority' (#40).[22] Yet this merely pushed the question back one stage. Granted the public, in the form of those who represented them, had authority to adjudicate on the kind of shared and contestable ethical and social issues generated by science's progress, on what (or whose) criteria were they to make their decisions? If a judgement were to be made about the limits on legal abortion, the lawfulness of IVF, the objectives of genetic modification (whether in crops or humans), the legality of physician-assisted suicide, or any of the other ethical and social issues over which there is legitimate public disagreement,[23] whose worldview—which ethical principles, systems, values, and objectives—should be taken into account, and which should be considered decisive?

Interestingly, this very question was played out before our eyes during the period of the research for this book. The first expert interviews were conducted in late 2019, and by the time we were speaking to our thirteenth, news about a possible pandemic was spreading around the world. Referring to this impending threat, interviewee 14, a science journalist, commented, 'I don't want to hear from local religious authorities about it because I don't think they have any relevant expertise. It seems that is purely a scientific matter' (#14). This was an early example of 'follow the science'. A dozen interviews (and a few months) later, a professor commented on how critical the role of science was in the on-going situation but added that 'the scientists don't agree on the SAGE [Scientific Advisory Group for Emergencies] committee', and that meant that 'the politicians listening to the discussion have still got to make a decision' (#26). By the end of the research, in early 2021, one interviewee with expertise in both science and the humanities could comment, in words some of which we have already quoted, that 'if we look at the Covid outbreak, the politicians kept telling us we're being led by the science, as though science could tell us whether it was better to protect ... physical health of old people [or] the mental health of young people' (#96). We can trace a path from 'science is all we need', through 'politics is *de facto* required on account of scientific uncertainty', to 'we *in principle* need other forms of reasoning because the science is not enough' through these three examples.

It would be a mistake to make too much of the sequence of just three interviews. It could well be (indeed, it is likely) that interviewee 96 was alert to science's limitations in late 2019 whilst interviewee 14 was still advocating 'follow the science' in mid-2021. But the pattern of the three quotations is interesting nonetheless, in the way it hints at a growing awareness, which may reflect wider public opinion during this period, that even issues as 'scientific' as a pandemic—with science playing an indispensable role in determining the make-up, contagiousness, spread, and treatment of the virus—were still amenable to, and indeed in need of, a non-scientific contribution. Responding to Covid meant not only decoding its genetic make-up, calculating infection rates, and designing an effective vaccine, but also thinking through the relative significance of, for example, lives, mental health, personal freedom, and economic security within our collective response. Contribute to this as it might, science does not have the authority to determine it.

244 THE LANDSCAPES OF SCIENCE AND RELIGION

Nor, of course, does religion, and here lies the real fault line in this part of the science and religion landscape, namely the basis, extent, and reasoning of religious contributions to debates on the contentious ethical and social issues thrown up by scientific progress. No interviewees—not even the most religiously hostile—were in favour of banning religious contributions to these debates. To do so would have been to mark oneself out as inexcusably illiberal. But many did frequently raise questions about the platform from which these religious contributions were made.

Religious authorities, they claimed, 'are given a kind of status when ethical questions come up' (#5). Religion is treated with an undeserved 'respect or separation' (#9). Establishment amplified the church's voice, positioning it 'at the head of a queue or ... in the legislature in ways that are very unbalanced relative to the commitment in the population at large' (#61). The oft-mentioned presence of bishops, by right, in the House of Lords epitomised this. 'The idea that there should be bishops in the House of Lords is grotesque. I don't understand why it still happens' (#46). Ultimately, if 'the Archbishop of Canterbury [is to] have a voice ... it should not be a power or authority [but] just another voice' (#24).

Those who made these points were sometimes at pains to stress that this was a principled argument rather than simply borne of personal frustrations. 'It so happens that we often have very good bishops in the House of Lords [but] that doesn't alter the principle' (#12). However, on occasion a mask would slip, such as when one science commentator said, 'the more I disagree with what the religion is saying, the less I want them to have power' (#9).

The question of religious status in these debates was closely linked to that of the reasoning underlying it. Critics were clear that just as religion shouldn't be afforded particular authority, nor should it be able to appeal to particular authority.

> If the premise of the view were that this is forbidden by the deity, or [that] 2,000 years ago that was the view that was taken by the founders of our movement, that kind of argument, *argumentum ad verecundiam*, the appeal to authority, isn't and shouldn't be [acceptable]. (#61)

Arguments that are based on unchallengeable authority, on revelation, on the ethics of a particular tradition, on venerated scriptures, these are not acceptable in contemporary public reasoning. In as far as religious participation in the public debates raised by science draws on any of these, it is, according to this argument, inadmissible.

PUBLIC AUTHORITY AND PUBLIC REASONING   245

This is a longstanding debate, closely associated with the arguments against religious contribution to public debate articulated by John Rawls in his *Theory of Justice*. Although no-one mentioned Rawls by name, in as far as interviewees defaulted to any position here, it was to the more exclusivist approach associated with Rawls' earlier work, which argued that reasoning had to appeal to political principles that were available and comprehensible to all reasonable citizens in order to be publicly admissible. In effect, if the religious were to contribute to this kind of public debate generated by scientific progress, they couldn't do so by drawing on any distinctively religious ideas, texts, or authorities.

As has been pointed out by numerous critics, this criterion of public reasoning makes an unsustainably arduous demand on particular sectors of the population. Rawls himself rowed back from it in his *Political Liberalism*, published in 1993, and even that attenuated position has subsequently come under challenge by political theorists.[24] Whether or not it is advisable (it usually is not), it should be *permissible* for any contributor to public debate, religious or not, to draw on their deepest convictions, even if they are not shared by or even comprehensible to 'all reasonable people'.

In reality, 'all reasonable people' is a rather question-begging concept, assuming a reasonable and consensual position on contentious ethical issues that is rarely there. The way in which one critic of religion told us 'the fact is nobody disagrees with the fact that life is sacred' (#15) underlines the point. Not only *do* people disagree with that view, but even many of those who are sympathetic towards it would demur from using a word like 'sacred' to defend their views. Assumptions of what 'all reasonable people' agree on are just that—assumptions. 'Reasonable people' all too often is a(n unconscious) cipher for 'people who think like me'.

This was underlined when one philosopher commented on this whole issue that 'religious people could make arguments that I'm going to find persuasive potentially, *if they appeal to stuff that I'm not sceptical about*' (#42; emphases added). On the surface, this sounds an obvious and eminently reasonable position. Of course, 'I am only going to be persuaded by "stuff" about which I'm not sceptical'. It is a basic approach to all rhetoric and argumentation that you should appeal to what your auditors find appealing.

However, when elevated to the level of public debate, particularly in a plural society like those of the contemporary West, this becomes much more problematic. Who is to say what 'stuff' is obvious and what 'stuff' dubious?

246 THE LANDSCAPES OF SCIENCE AND RELIGION

Why should one particular individual's underlying beliefs and commitments be judged the only legitimate currency of public reasoning? Why should the 'the stuff that I am not sceptical about' be the yardstick for what is deemed reasonable and admissible in public debate? It is surely more reasonable and realistic to expect dissensus, even among reasonable people (whoever they may be) than consensus? Ultimately, to deny some but not others access to their deepest convictions in public reasoning is not to treat all participatory citizens equally.[25]

The critical point in all this is to highlight how it is this issue that constitutes the relevant front to this part of the landscape. When people are disagreeing about science and religion in relation to social and ethical issues, it is only superficially about the issues themselves. Deeper down lie questions of different respective stances towards the world—progressive or conservative, for example. But deeper even than these lie questions of public authority and reasoning, meaning who gets to decide on which respective stances towards the world prevail, and why.

Simply because science is making so many advances—such as about the start and end of life, concerning our genetic makeup and ability to correct, modify, and 'improve' nature, and may well make further progress on postponing death, advancing AI, or finding moral awareness and consciousness in non-human animals—it is going to open up questions that touch on the deepest anthropological considerations, as explored in the previous chapter.[26] That being so, it will catalyse public debates that draw in this deep reasoning, and in the process open up the question of whose contributions to these debates matter, and why. And while this is not in any way a narrowly or uniquely science and religion issue, it is and will be one that occurs against a science and religion backdrop, and will therefore undoubtedly be seen, by some, as another thing people are disagreeing about when they disagree about science and religion.

Outstanding questions

What that comes down to, I suppose, is a political power struggle between people of progressive, more secular bents as opposed to those of a more conservative, and conservatively religious bent. But in a sense, these are socio-political tensions that happen to be characterised by religious labels … where the disagreements are clearly very coloured by a particular religious … and a non-religious world view. All that again is just a long-winded way of saying I don't think that there's any, in principle, incompatibility here. It's just that you happen to have religion still serving as a form of social divide. (#28)

In the popular mind, and not only the popular mind, science and religion disagree, sometimes severely and irreconcilably, over a range of contentious ethical and social issues. Science says we can. Religion says we shouldn't. The results are played out in the headlines.

On closer inspection, however, this is not—or at least need not be—a narrowly science and religion conflict. Quite apart from anything else, the role of the media in all this was roundly criticising, not simply for exaggerating conflict but (especially by scientists) for simplifying science.[27] More pertinently, neither religion, with its more complex attitude to the modification of the material world than is captured by the conservative/archaic caricature, nor science, with its widely acknowledged need for ethical evaluation and limitation alongside its naturally progressive character, was up to playing the role of straightforward antagonist in a conflict between progressive and conservative stances towards the world. Science needs a broader ethical context and constraint, and the religious voice could, quite properly, contribute to that.

But that voice was one among many, and the deeper tension lay in the much broader and long-standing question of which voices, which traditions, and which institutions should contribute and/or have authority in these shared and contested debates and, alongside it, which arguments, reasons, values, and commitments should hold sway. As the sociologist quoted above implies, this is not in principle a science and religion issue. Indeed, it is one with which every plural society that lays claim to liberal principles will necessarily wrestle. But it is an issue that does naturally don the 'colours' of science and religion and play out on the science and religion landscape, and it is an issue that is seen by a watching crowd as one more thing that science and religion disagree about.

Notes

1. Noy, S. and O'Brien, T.L. 'A nation divided: science, religion, and public opinion in the United States'. *Socius* (2016), 2, 1–15.
2. These three categories capture three broad perspectives on science and religion in US opinion. 'Moderns' hold science in relatively high and religion in relatively low esteem; 'traditionals' hold the reverse; and the 'postsecular' position 'is knowledgeable about and appreciative of science but ... is religiously devout and ... rejects mainstream scientific

248 THE LANDSCAPES OF SCIENCE AND RELIGION

accounts of evolution and the big bang'. The positions are derived from and laid out in greater detail in O'Brien, T.L. and Noy, Shiri, 'Traditional, modern, and post-secular perspectives on science and religion in the United States', *American Sociological Review* (2015), 80(1), 92–115.

3. The group they class as 'moderns', for example, hold more progressive views about gender roles, sexuality, pornography, and sex education but are less supportive of affirmative action when it comes to race and civil liberties. By contrast, 'traditionals', in addition to being less supportive of abortion rights or of making contraceptives accessible to teenagers, are more likely to believe that success in life is due to internal factors (e.g. hard work) rather than external ones (e.g. luck or help). Not all possible differences were visible between these groups. 'Religious–scientific perspectives do little to differentiate public opinion about government and its role in citizens' lives', for example, with 'moderns', 'traditionals', and 'postseculars' holding broadly similar views on social assistance programmes. Nevertheless, the correlation between what someone thought about science and religion and their attitude to such a wide range of socio-economic and cultural issues was clearly not co-incidence.

4. O'Brien and Noy, 'Political identity and confidence', pp. 439–461.

5. Elsdon-Baker, Fern, et al., *Science and Religion: Exploring the Spectrum: A Multi-Country Study on Public Perceptions of Evolution, Religion and Science*, University of Birmingham/YouGov (December 2023), UoB-YouGov.-Science-and-Religion-Survey-Report.-8-Dec-2023-.pdf (scienceandbeliefinsociety.org).

6. 'I do see an affinity [between] being non-religious and a sense of a right to decide your own death, but I don't see that feeding in very significantly into ideas about the religious ... science has this ... sort of idea of scientistic non-religious culture [which] is more associated with rationalist humanism than it is even with other forms of humanism that focus much more on the social sides of humanity.' (#30)

7. The US Supreme Court's overturning of *Roe v. Wade* occurred after our fieldwork was finished, but its decision provoked fears of a religiously based (or at least religiously linked) limit of IVF (https://www.theguardian.com/us-news/2022/may/12/ivf-treatment-us-anti-abortion-laws-bills).

8. 'One reason why ... African Americans at the moment will be dubious about the [Covid] vaccine is that they were experimented upon by white scientists way back in this century.' (#73)

9. The full list was (1) Using and then destroying human embryos if doing so helps scientists find cures for disease; (2) Creating a baby that is a clone of another person; (3) Using genetic engineering to create a baby that is smarter, physically stronger, or better looking; (4) Medical research that uses stem cells from sources that do *not* involve human embryos; (5) Use of reproductive technologies to identify diseases in the womb; (6) Euthanasia or physician-assisted suicide; (7) Using IVF: when an egg is fertilized by sperm in a test tube or elsewhere outside the body); (8) Having an abortion for any reason; and (9) Having an abortion if there is a strong chance of serious defect in the baby.

10. This was not invariably the case. For example, when asked whether they 'would prefer a robot to conduct surgery on me than a human surgeon', 13% of biblical literalists agreed compared to around 10% of those with a higher science qualification. That said, this is only a small difference and the sample size of the first group (172) means that the result should be treated with considerable caution.

11. 'The moment of creation of life is a God-given moment and we shouldn't do that. Clearly the fact that we can do that I think negates that argument.' (#8)

12. 'There's some God-given difference between human beings and other things ... which means that you can't then abort foetuses or design babies, or whatever it might happen to be. At that point, they've lost me.' (#42)

13. 'The Church of England actually allowed people who were trans to marry before they allowed gay people to marry. They're actually quite liberal in their attitude towards trans individuals ... if you look at Hinduism and Buddhism, they've never categorised people as

existing within two genders. Hinduism has three. Buddhism has four, so this notion that all religions hold these archaic Christian views is just a fabrication by the media.' (#99)

14. 'Maybe I have a scientific belief ... [but] in practice that does diddley for me when I have to walk past homeless people at my station.' (#7) 'But physics is not terribly useful for advising me on how to be a good parent.' (#15)

15. 'When I was a scientist really ... my entire universe was about this one molecule called Klp5 and yet it was all-consuming, and you just do not have the time or the headspace really to dedicate to looking at the sort of long-term implications or the ways in which this could be used.' (#74)

16. In a similar vein, 'one of the reasons why scientists need to be really careful when they're in the public eye ... if they are giving what are essentially personal views ... it comes with this halo of clarity that comes from their job.' (#95)

17. 'For example ... [when] you try to decide whether the National Health system of blood donation, or the US system, 30 years ago, being paid for, was better ... medical data are very useful and we learn a lot about likelihood of getting things like HIV and hepatitis in both systems, but we also learn about shortage of blood for particular operations but still ... that won't solve the ethical arguments.' (#11)

18. 'I think religious people can have an input in science. Not because they are religious, just because they are not scientists.' (#29)

19. The interviewee had in mind the just war tradition in this instance.

20. This reminds us of the irregular verbs that the civil servant Bernard Woolley mentioned in the BBC comedy *Yes, Minister*: 'I give confidential security briefings. You leak. He has been charged under section 2a of the Official Secrets Act.' In this instance, it is perhaps: 'I suggest, you demand, he imposes.'

21. This is one of the core arguments of Spencer's *Magisteria*.

22. The picture is, of course, more complex than this, as the same interviewee went on to acknowledge. 'I realise that there are other unelected people who have authority, people like judges for example.'

23. And indeed, over the background question of whether public disagreement over an issue is indeed legitimate—a question that is raised in wider discussions concerning the extent and negotiability of various human rights.

24. See, for example, Chaplin, Jonathan, *Talking God: The Legitimacy of Religious Public Reasoning* (London, 2008).

25. Of course, whether it is advisable for such contributors to draw openly on those convictions, certainly without going through the hard work of translating them into a more accessible register (e.g. the rendering of 'sanctity' to 'dignity' in much religious public reasoning), is another matter altogether.

26. All these topics are explored in our book *Playing God: Science, Religion and the Future of Humanity* (London, 2024).

27. 'The way that media then translates a scientific paper is really problematic, they often give it far more weight than the scientists themselves claim if you go back into the original papers.' (#85) 'For a lot of scientists, their first foray into the media is quite bruising ... scientists who've been doing media, work for years, have taken years to get over that bruising, but they can still tell you how bruising it is.' (#95) 'I think it misleads people and you wind up with an almost cartoon version of science, where either it's this truth handed down from on high that you're just supposed to accept [...].' (#35) There were other similar comments. In all fairness, while there were quite a few science journalists and commentators in amongst the interviews, the case for the defence—how else do you communicate complex scientific ideas to a general public?—was not given much of an airing.

Conclusion

What are people disagreeing about when they disagree about science and religion? Or, put more precisely, when people think that science and religion are in tension, in disagreement, or in wholesale conflict with one another, what exactly do they think they are differing over?

The question can be framed in more positive terms. When people argue that science and religion are not in conflict, on what basis do they do so? This formulation seems odd, but, on closer inspection, it not only makes sense but also can be rather fruitful. 'A' may think that science and religion are in harmony on the basis that science has proved true all the dogmatic claims made in a particular religion's confession or creed. 'B' might claim concord on the grounds that the two entities are really non-overlapping magisteria, with one's interest in values having no bearing on the other's concern for facts. 'C' could maintain there is agreement between them because science is, at heart, no more than a particular, structured methodology for interrogating the material world, the practice of which poses no threat to a religion's metaphysical beliefs. And 'D' may assert there is peace on the basis that, properly speaking, religion is no more than the ritualised practice of communal belonging, making no claims about reality with which science does or even could disagree. For each, harmony is asserted on the basis of a particular prior understanding of the category in question, the resulting consonance, to quote one of our epigraphs, being no more than 'a contrived artefact'.

Approached this way, the importance of understanding what exactly is being compared becomes clear. To quote from the same epigraph, unless we can 'get inside the language games [or] paradigms of ... respective traditions', then any model we build concerning their relationship will be constructed on fragile foundations.

This book has tried to advance that understanding by exploring science and religion as concrete contemporary 'practices', attempting to grasp what people today, both experts and the general public, think science and religion are. The result has been two 'landscapes', sprawling, ill-defined, contiguous at various points, sometimes even partially overlying one another, and replete with various common features—'landscape' being a metaphor that has stood for the expansive, imprecise, disputable, socially constructed categories that 'science' and 'religion' actually are.

Each landscape is recognisable not on account of any clear or officially regulated boundary around it but simply from the number of relevant features discernible within it. In effect, the more a particular human activity is characterised by the kind of features we discerned in the various legal, philosophical, and sociological discussions of science—features concerning method, object of study, presuppositions, objectives, values, and institutional structure—the more likely that activity can legitimately be called 'science'.

In a similar way, the more a human activity is characterised by a concern for the transcendent, by a particular supportive attitude to belief, by personal commitment, and by the formalisation and institutionalisation of both that belief and that commitment, the more likely it can legitimately be called 'religion'. If the resulting definitions are messy and fuzzy, that is simply the cost of getting inside their respective language games (or paradigms), of seeking to understand science and religion as they are, as concrete contemporary 'practices'.

Such an exercise could serve as a means of understanding why people believe the two to be compatible, mapping out the various ways in which the two landscapes merge peacefully or simply co-exist harmoniously. For much of the time and for many people, there is no conflict to be seen here. Indeed, for many people there is active coherence and complementarity between the two. Science is characterised by commitment to certain metaphysical presuppositions that are well aligned with, even justified by, certain religious doctrines. Science requires ethical commitments of the kind that are advocated in religious texts and formalisations. Religion is characterised by the institutionalisation of practices and offices that finds more than an occasional echo in forms of scientific institutionalisation. There is indeed a great deal of consonance between the concrete contemporary 'practices' of each.

However, if we are honest, this is not really the question that motivates most discussion in this field. The world of science and religion—with its now-discredited but stubbornly popular history of conflict, its resurgent anti-evolutionary 'fundamentalist' views, its loud and antagonistic scientific provocateurs, and its default relationship, at least among UK adults, of perceived disharmony—is characterised by the perception of conflict. The obvious question to ask is why.

Through these landscapes, we have identified four issues in particular around which disagreements congregate. Whether it is over the metaphysics, methodology, anthropology, or attitudes to public authority and reasoning, people do disagree about science and religion.

It is not always the same people. Following the lead of John Evans, we agree that *popular* tensions are more likely to fixate on the perceived ethical and social 'content' of the two (in particular, when the question of the nature and value of the human comes into view) and also on the question of which entity should have the right to contribute to and adjudicate on shared issues of public contention. By contrast, expert opinion is more likely to fixate on the perceived difference between the metaphysical commitments and methodological approaches of each. But this is a rule of thumb rather than an iron law, and much of the popular sense of conflict is rooted, deep down, in metaphysical and methodological considerations (albeit rarely expressed in those terms), while experts are hardly blind or indifferent to any wider social and ethical tensions.

Many of those disagreement, we have argued, are thoroughly negotiable. Time and again, the experts to whom we spoke recognised that for all that some claimed there was an insurmountable conflict between science and religion over their respective metaphysical commitments, approaches to acquiring and retaining knowledge, conceptualisations of the human, and belief in who has the right to inform and decide contentious social and ethical issues, in actual fact, that conflict could be obviated. In effect, much of the time when people disagreed about science and religion, they needn't be doing so.

But much of the time is not all the time, and we have also argued that areas of tension—what we have called 'outstanding questions'—remain. Some relate to metaphysics—such as whether methodological naturalism demands a commitment to strict ontological naturalism, or how necessary is complete closed causality to the functioning of science, or how inimical is it to the religious concern for the transcendent? Some relate to methodology—such as how far can revelation be considered a legitimate source of information, how much weight can be placed on subjective experience as evidence for truth claims, or how is theological consensus arrived at and on what basis?

Some relate to the content of each, in particular anthropological content—such as how coherent and how watertight is the category of the 'human', or how can a religious doctrine of creation and love be squared with the sheer scale and apparent necessity of pain and suffering as revealed by science, now that recourse to an actual Fall and pre-Lapsarian past is no longer possible? And some, finally, relate to the question of public authority and public reasoning—such as who is to balance the appropriate roles of religious and scientific institutions, leaders, and arguments in the adjudication of contentious public ethical issues?

Some of these questions will be familiar in other guises, e.g. the problem of miracles, the problem of suffering. Some, such as the question of authority in a plural public square, are hardly specific to science and religion. And some, such as the kind of naturalism to which science must and religion can be committed, are unlikely to keep anyone other than professional philosophers and theologians awake at night. But they are all, we believe, interesting and important issues, and worthy of further discussion.

Appendices

Appendix 1 Qualitative research

For the interviews that lie at the heart of this book, and in particular chapters 2, 4, and 6–9, we spoke to 101 interviewees from a wide range of backgrounds within science, philosophy, religion, and communication. Interviews lasted around an hour, although some were (much) longer and a few (slightly) shorter. Interviews were conducted between the autumn of 2019 and spring of 2021. All were recorded and transcribed for analysis. Although many interviewees said they were happy to be quoted in person, a few were not, and we have preserved the anonymity of all quotations (interviewees were told that they were going to be named as interviews but not identified further; the numbers for interviewees in the text do not correspond with the list below, which is alphabetical).

Interviewees were recruited primarily for their professional expertise, though we were clear that we wanted a majority of non-religious and non-believing interviewees, primarily because we were interested in locating the perceived tension points between science and religion and this is easier to do with non-religious and non-believing respondents. We did not, however, recruit on the basis of their views of science and religion compatibility, which we only discovered after inviting them to interview.

Of our interviewees, 60% were male and 40% female; 63% were not religious and 31% were (6% did not say). When it came to their stated beliefs, 55% said 'I do not believe in God'; 6% said 'I don't know whether there is a God and I don't believe there is any way to find out'; 7% said 'I don't believe in a personal God, but I do believe in a higher Power of some kind'; 6% said 'I find myself believing in God some of the time, but not at others'; 15% said 'While I have doubts, I feel that I do believe in God'; and 13% said 'I know God really exists and I have no doubts about it'.

When it came to their attitude to science and religion compatibility, 12% said they thought the two were strongly incompatible, 24% that they were incompatible, 41% that they were compatible, and 23% that they were strongly compatible. This came as something of a surprise to us—their attitudes were somewhat more compatible than the general public who are, on balance, more religious and more believing than this group of interviewees.

Interviews were conducted by means of a semi-structured interview guide, and were recorded. All the interviews were transcribed in full (resulting in close to a million words worth of transcripts) and then analysed by means of detailed re-reading and coding, using NVivo software. A series of meta-categories were then identified—e.g. physical sciences, life sciences, brain sciences, epistemology, metaphysics, methodology, religion, science and religion models, conceptions of the

APPENDICES   255

human, ethical discussions, and social and political discussions. These were then further broken down into sub-categories, with the connections between each being drawn out.

A full list of interviewees, given in alphabetical order and not indicative of the ascriptions within the report, follows.

- David Adam, science journalist
- Joshua Andrews, Lecturer in Eastern Religion, with special interest in ethics and existentialism (University of Bangor)
- Bryan Appleyard, award-winning science journalist and author
- Hana Ayoob, science communicator, podcast host, and speaker
- David Baddiel, author, screenwriter, television presenter, and playwright
- Julian Baggini, philosopher, journalist, and author
- Philip Ball, science journalist and author
- Helen Beebee, Samuel Hall Professor of Philosophy (University of Manchester) and President of the British Society for the Philosophy of Science
- Michael Berry, Professor of Mathematical Physics (University of Bristol)
- Sue Black, Professor of Anthropology (Lancaster University), President of the Royal Anthropological Institute of Great Britain and Ireland, and popular author
- Sue Blackmore, parapsychologist and author
- Paul Braterman, Professor of Chemistry (University of North Texas and University of Glasgow), science writer and educator
- Justin Brierley, broadcaster and podcast host
- Andrew Brown, journalist, religion correspondent, and editor
- Steve Bruce, Chair of Sociology (University of Aberdeen) and contributor to the British Social Attitudes Survey
- Geoffrey Cantor, Professor of the History and Philosophy of Science (University of Leeds)
- Bernard Carr, Professor of Mathematics and Astronomy (Queen Mary University of London)
- Tom Chivers, science journalist
- Stewart Clark, Professor of Physics (University of Durham)
- Stuart Clark, author, astronomer, broadcaster, consultant for the European Space Agency, and former editor of *Space Science*
- Joanna Collicutt, Karl Jaspers Lecturer in Psychology and Spirituality (University of Oxford)
- Brian Cox, Professor of Particle Physics (University of Manchester), Royal Society Professor for Public Engagement in Science, author, broadcaster, TV host, and documentary maker
- Lee Cronin, Regius Chair of Chemistry and Director of the Cronin Group (University of Glasgow)
- James Crossley, Professor of Bible, Society and Politics (St Mary's University, London)

256 APPENDICES

- Celia Deane-Drummond, Professor of Theology and Science (Durham University), Research Fellow in Theology (University of Oxford), and Director of the Laudato Si Research Institute
- Chris Done, Professor of Astrophysics and Theoretical Physics (University of Durham)
- Sarah Dry, author and historian of science
- Robin Dunbar, Professor of Evolutionary Psychology (University of Oxford) and author
- Fiona Ellis, Professor of Philosophy, and Director for the Centre of Philosophy for Religion (University of Roehampton, London)
- Miguel Farias, Reader in Cognitive and Biological Psychology (Coventry University)
- Myriam Francois, broadcaster, writer, and documentary maker
- Clive Gamble, Professor of Archaeology (University of Southampton), Fellow of the British Academy, and author
- Rose George, author and science journalist
- Robin Gill, priest, theologian, and ethicist
- Philip Goff, Philosopher (University of Durham) and author
- John Gowlett, Professor of Archaeology, Classics and Egyptology (University of Liverpool)
- A.C. Grayling, Master of the New College of the Humanities and Professor of Philosophy (Northeastern University London), and author
- Susan Greenfield, Baroness, Research Scientist (University of Oxford), author, and broadcaster
- Wendy Grossman, science and technology journalist
- Sarah Harper, Director and Core Professor of Gerontology and Founding Director of the Oxford Institute of Population Ageing (University of Oxford)
- Mark Harris, Professor of Natural Science and Theology (University of Edinburgh)
- Brenna Hasset, Archaeology Researcher (University College London), author and public speaker
- Victoria Herridge, Evolutionary Biologist and Scientific Associate at the Natural History Museum
- Roger Highfield, science director at the Science Museum Group, science journalist and broadcaster
- Richard Holloway, former Bishop of Edinburgh, author, and broadcaster
- Rowan Hooper, Head of Features at *New Scientist*, biologist, and science writer
- Isabella Kasselstrand, Lecturer in Sociology of Religion (University of Aberdeen)
- Eleanor Knox, Reader in Philosophy (Kings College London)
- Stephen Law, philosopher and author
- Graham Lawton, science journalist for *New Scientist*
- Sally Le Page, YouTuber and science communicator
- Lois Lee, Research Fellow in Religious Studies (University of Kent)

APPENDICES   257

- Joanna Leidenhag, Lecturer in Science-Engaged Theology (University of St. Andrews)
- John Lennox, Professor of Mathematics (University of Oxford), author, and renowned speaker on science and religion
- Beth Lord, Professor and Head of Philosophy (University of Aberdeen), and editor and executive committee member of the Society for European Philosophy
- Jo Marchant, science journalist and author
- Michael C. Marshall, science journalist and author
- Katherine Mathieson, CEO of the British Science Association
- Iain McGilchrist, author and psychiatrist
- Robin McKie, science journalist, author, and science editor for *The Observer*
- Felicity Mellor, Lecturer in Science Communication and Science Journalism (Imperial College London)
- Zeeya Merali, science journalist (writes for *New Scientist* and *Scientific America*)
- Neil Messer, Professor of Theology (University of Winchester) and author of books on Christian ethics, and science and religion
- Simon Mitton, astronomer, writer, and former editor
- Richard Norman, Professor of Moral Philosophy (University of Kent), Vice-President of Humanists UK, and founding member of the Humanists Philosophers' Group and Humanist Peace Forum
- Kelly Oakes, science journalist
- David Papineau, Professor of Philosophy of Science (Kings College London)
- John Parker, Head of Department, Mathematical Science, Professor of Geometry (University of Durham)
- Sumit Paul-Choudhury, journalist and former editor-in-chief at *New Scientist*
- Arthur Petersen, Professor of Science, Technology and Public Policy (University College London), editor of *Zygon: Journal of Religion and Science*, and visiting Professor at Massachusetts Institute of Technology and London School of Economics
- Duncan Pritchard, Professor of Philosophy (University of Edinburgh)
- Yakub Qureshi, editor at Reach PLC and journalist
- Gina Radford, previously Deputy Chief Medical Officer for England, reverend in Church of England, and guided church response to COVID-19
- Amanda Rees, Lecturer in Sociology, Science and History (University of York)
- Martin Rees, Professor of Cosmology and Astrophysics (University of Cambridge), former President of the Royal Society, and member of the House of Lords
- Michael Reiss, Professor of Science Education (University College London)
- Kathleen Richardson, Professor of Ethics and Culture of Robots and AI (De Montfort University) and author
- Sarah Lane Ritchie, Lecturer in Science and Religion (University of Edinburgh)
- Adam Rutherford, Lecturer in Genetics, author and broadcaster

258 APPENDICES

- Angel Saini, author, broadcaster, journalist, and documentarian
- Tasia Scrutton, Associate Professor of the Philosophy of Religion (University of Leeds)
- Alom Shaha, scientist and author of *The Young Atheist's Handbook*
- Rupert Sheldrake, author and biologist
- Hayaatun Sillem, CEO of the Royal Academy of Engineering
- Charlotte Sleigh, Honorary Professor of Science, Humanities and History (University of Kent), Honorary Professor of University College London, and President of the British Society for the History of Science
- Bethany Sollereder, Fellow in Science and Religion (University of Oxford)
- Chris Southgate, Professor of Theology and Religion (University of Exeter)
- Francesca Stavarakopoulou, Professor of Hebrew Bible and Ancient Religion (University of Exeter) and broadcaster
- Chris Stringer, Professor and Research Leader in Human Evolution (Natural History Museum) and author
- Adrian Sutton, Professor of Natural Sciences (Imperial College London)
- John Swinton, Professor of Practical Theology and Pastoral Care, Master of Christ's College, and Director at the Centre for Spirituality, Health and Disability (University of Aberdeen)
- Amelia Tait, freelance journalist
- Ray Tallis, philosopher, author, physician, and clinical scientist
- Amy Unsworth, Research Fellow in Science and Religion (University College London)
- David Voas, Professor and Department Head of Social Science (University College London), and Contributor to the British Social Attitudes Survey
- Martin Ward, Temple Chevallier Chair of Astronomy (University of Durham) and previous consultant to the European Space Agency
- Tom Whipple, science editor at *The Times*
- David Wilkinson, Professor and Principal of St. John's College (University of Durham), scientist, and theologian
- Richard Wiseman, Professor of Public Understanding of Psychology (University of Hertfordshire), author, and lecturer
- Linda Woodhead, Professor of Sociology of Religion (Lancaster University)
- Rebecca Wragg Sykes, author and archaeologist (University of Liverpool)

Appendix 2 Quantitative research

The quantitative element of this research, which underpins chapter 5 in particular, surveyed 5,153 UK adults in fieldwork conducted by YouGov between 5 May and 13 June 2021. The survey was conducted using an online interview administered to members of the YouGov Plc UK panel of 800,000+ individuals who agreed to take part in surveys.

Emails were sent to panellists selected at random from the base sample. The email invited them to take part in a survey and provided a generic survey link. Once a panel member clicked on the link, they were sent to the survey that they were most required for, according to the sample definition and quotas. Invitations to surveys did not expire and respondents could be sent to any available survey. The responding sample was weighted to the profile of the sample definition to provide a representative reporting sample. (The profile was normally derived from census data or, if not available from the census, from industry-accepted data.)

The questionnaire for the survey is available at www.theosthinktank.co.uk.

Bibliography

Legal cases

- *Scientific Societies Act* (1843)
- *Davis v. Beason* (1890)
- *United States v. Macintosh* (1931)
- *United States v. Ballard* (1944)
- *Fowler v. Rhode Island* (1953)
- *Torcaso v. Watkins* (1961)
- *United States v. Seeger* (1965)
- *Welsh v. United States* (1970)
- *R v. Registrar General, ex parte Segerdal* (1970)
- *Malnak v. Yogi* (1979)
- *International Society for Krishna Consciousness, Inc., v. Barber* (1981)
- *Rev. Bill McLean v. The Arkansas Board of Education* (1982)
- *Edwin W. Edwards, Governor of Louisiana v. Don Aguillard* (1987)
- *Transport Museum Society of Ireland Ltd. v. Registrar of Friendly Societies* (1999)
- *Kitzmiller v. Dover Area School District* (2005)
- *R (on the Application of Hodkin) v. Registrar General of Births, Deaths and Marriages* (2013)

Data sources

- Theos/Faraday/YouGov, available at www.theosthinktank.co.uk
- Public attitudes to science: GOV.UK (www.gov.uk)
- Education, England and Wales: Office for National Statistics (www.ons.gov.uk)
- Higher Education Student Statistics, UK, 2020/21—subjects studied: HESA
- Improving the fortunes of the humanities means thinking about post-16 qualifications: HEPI
- The Global Faith and Media Study: Media Diversity Institute (www.mediadiversity.org)
- Americans' Trust in Scientists, Other Groups Declines: Pew Research Center
- Public Perception of Genetics: Genetics Society
- Science and Religion. Exploring the Spectrum: A Multi-Country Study on Public Perceptions of Evolution, Religion and Science: University of Birmingham/YouGov, December 2023, UoB-YouGov.-Science-and-Religion-Survey-Report.-8-Dec-2023-.pdf (scienceandbeliefinsociety.org)
- Pew Forum: Science and religion in central and eastern Europe: Pew Research Center (www.pewforum.org)

Books and articles

Abrahams, William, 'The offense of divine revelation', *Harvard Theological Review* (2002), 95(3), 251–264

Alexander, Eben, *Proof of Heaven: A Neurosurgeon's Journey into the Afterlife* (London, 2012)

Alston, William, *The Philosophy of Language* (Englewood Cliffs, NJ, 1964), p. 90

Asma, Stephen, *Why We Need Religion* (New York, 2018)

Babbage, Charles, *Reflections on the Decline of Science in England* (Farnborough, 1969 [1830])

Baker, Joseph O. 'Public perceptions of incompatibility between "science and religion"', *Public Understanding of Science* (2012), 21, 340–353

Barbour, Ian G., *When Science Meets Religion: Enemies, Strangers, or Partners?* (San Fransisco, 2000)

Bardon, Aurélia and Howard, Jeffrey William (eds.), *Liberalism's Religion: Cécile Laborde and Her Critics* (London, 2020)

Bauer, Henry 'Science in the 21st century: knowledge monopolies and research cartels', *Journal of Scientific Exploration* (2004), 18(4), 643–660

Bellah, Robert, *Religion in Human Evolution* (Cambridge, MA, 2011)

Berger, P.L., *The Sacred Canopy* (Garden City, NY, 1967)

Berger, P.L., 'Some second thoughts on substantive versus functional definitions of religion', *Journal for the Scientific Study of Religion* (1974), 13, 125–133

Blakely, Jason, We Built Reality: How Social Science Infiltrated Culture, Politics, and Power (Oxford, 2020)

Brigham and Women's Hospital, 'Brain circuit for spirituality?', *Science Daily* (1 July 2021)

Brooke, John Hedley, *Science and Religion: Some Historical Perspectives* (Cambridge, 1991)

Brooke, John Hedley and Numbers, Ronald L. (eds.), *Science and Religion around the World* (Oxford, 2011)

Chan, E., 'Are the religious suspicious of science? Investigating religiosity, religious context, and orientations towards science', *Public Understanding of Science* (2018), 27, 967–984

Chaplin, Jonathan, *Talking God: The Legitimacy of Religious Public Reasoning* (London, 2008)

Chevassus-au-Louis, Nicolas, *Fraud in the Lab: The High Stakes of Scientific Research* (Cambridge, MA, 2019)

Collins, Francis S., *The Language of God: A Scientist Presents Evidence for Belief* (New York, 2006)

Coyne, Jerry A., *Faith Versus Fact: Why Science and Religion Are Incompatible* (New York, 2015)

Dawkins, Richard C., *The God Delusion* (London, 2006)

Deane-Drummond, Celia, *Christ and Evolution: Wonder and Wisdom* (London, 2009)

De Waal, Frans, *The Bonobo and the Atheist: In Search of Humanism among the Primates* (New York, 2013)

Dixon, Thomas, Cantor, Geoffrey, and Pumfrey, Stephen (eds.), *Science and Religion: New Historical Perspectives* (Cambridge, 2010)

Dubuisson, D., *The Western Construction of Religion: Myths, Knowledge, and Ideology* (Baltimore, 2003)

Duhem, Pierre, *Aim and Structure of Physical Theory* (New York, 1962)

Dupré, John, *The Disorder of Things: Metaphysical Foundations of the Disunity of Science* (Cambridge, MA, 1993)

262 BIBLIOGRAPHY

Durkheim, Émile, *The Elementary Forms of Religious Life* (Oxford, 2001 [1915])

Ecklund, Elaine Howard, *Science vs. Religion: What Scientists Really Think* (Oxford, 2010)

Ecklund, Elaine Howard and Scheitle, Christopher P., *Religion vs. Science: What Religious People Really Think* (Oxford, 2017)

Eliade, Mircea, *The Myth of the Eternal Return: Cosmos and History* (Princeton, NJ, 1971)

Elsden-Baker, F. 'Creating creationists: the influence of "issues framing" on our understanding of public perceptions of clash narratives between evolutionary science and belief', *Public Understanding of Science* (2015), 24, 422–439

Elson-Baker, Fern and Lightman, Bernard, *Identity in a Secular Age: Science, Religion and Public Perceptions* (Pittsburgh, 2020)

Evans, John, *Morals Not Knowledge: Recasting the Contemporary U.S. Conflict between Religion and Science* (Oakland, CA, 2018)

Evans, John H. and Evans, Michael S., 'Religion and science: beyond the epistemological conflict narrative', *Annual Review of Sociology* (2008), 34, 87–105

Falk, Dan, 'Learning to live in Steven Weinberg's pointless universe', *Scientific American* (27 July 2021)

Ferguson, Michael A., Schaper, Frederic LWVJ, Cohen, Alexander, et al., 'A neural circuit for spirituality and religiosity derived from patients with brain lesions', *Biological Psychiatry* (29 June 2021)

Flanagan, Owen, *The Bodhisattva's Brain: Buddhism Naturalized* (Cambridge, MA, 2013)

Frazer, James George, *The Golden Bough: A Study in Magic and Religion* (New York, 1925 [1890])

Freeman, George C., 'The misguided search for the constitutional definition of "religion"', *Georgetown Law Journal* (1983) 71(6), 1519–1566

Geertz, Clifford, *The Interpretation of Cultures* (New York, 1973)

Gingras, Yves, *Science and Religion: An Impossible Dialogue* (Cambridge, 2017)

Glennan, Stuart, 'Whose science and whose religion? Reflections on the relations between scientific and religious worldviews', *Science & Education* (2009), 18, 797–812

Gordin, Michael D., *On the Fringe: Where Science Meets Pseudoscience* (Oxford, 2021)

Gould, S.J. 'Nonoverlapping magisteria', *Natural History* (1997), 106, 16–25

Gould, S.J., *Rocks of Ages: Science and Religion in the Fullness of Life* (New York, 1999)

Gould, S.J., *The Hedgehog, the Fox, and the Magister's Pox: Mending the Gap between Science and the Humanities* (London, 2003)

Gregersen, N.H. and van Huyssteen, J.W. (eds.), *Rethinking Theology and Science: Six Models for the Current Dialogue* (Grand Rapids, MI, 1998)

Hardin, Jeff, Numbers, Ronald, and Binzley, Ronald A. (eds.), *The Warfare between Science and Religion: The Idea that Wouldn't Die* (Baltimore, 2018)

Harrison, Peter, *The Fall of Man and the Foundations of Science* (Cambridge, 2007)

Harrison, Peter *The Territories of Science and Religion* (Chicago, 2015)

Harrison, Peter '"I believe because it is absurd": the enlightenment invention of Tertullian's *Credo*', *Church History* (2017), 86(2), 339–364

Hart, David Bentley, *The Experience of God: Being, Consciousness, Bliss* (London, 2013)

Haught, J.F., *Science and Religion: From Conflict to Conversation* (New York, 1995)

Hunter, James Davison and Nedelisky, Paul, *Science and the Good: The Tragic Quest for the Foundations of Morality* (New Haven, 2018)

James, William, *Varieties of Religious Experience* (London, 1902)

Jong, Jonathan, 'On (not) defining (non)religion', Science, *Religion and Culture* (2015), 2(3), 15–24

BIBLIOGRAPHY   263

Kroesbergen, H., 'An absolute distinction between faith and science: contrast without compartmentalization', *Zygon* (2018), 53, 9–28

Laborde, Cécile, *Liberalism's Religion* (Cambridge, MA, 2017)

Labron, Tim, *Science and Religion in Wittgenstein's Fly Bottle* (New York, 2017)

Larson, Edward J., *Summer for the Gods: The Scopes Trial and America's Continuing Debate over Science and Religion* (New York, 2008)

Laudan, Larry, 'The Demise of the Demarcation Problem', in Cohen, R.S. and Laudan, L. (eds.), *Physics, Philosophy and Psychoanalysis: Essays in Honor of Adolf Grünbaum* (Dordrecht, 1983), pp. 111–127

Lightman, Bernard (ed.), *Rethinking History, Science, and Religion: An Exploration of Conflict and the Complexity Principle* (Pittsburgh, 2019)

Lindberg, David C. and Numbers, Ronald L. (eds.), *God and Nature: Historical Essays on the Encounter between Christianity and Science* (Berkeley and London, 1986)

McCauley, Robert, *Why Religion Is Natural and Science Is Not* (Oxford, 2013)

McFarland, Ian, 'Conflict and compatibility: some thoughts on the relationship between science and religion', *Modern Theology* (2003), 19, 181–202

McGilchrist, Iain, *The Master and His Emissary: The Divided Brain and the Making of the Western World* (New Haven and London, 2009)

McGilchrist, Iain, *The Matter with Things: Our Brains, Our Delusions, and the Unmaking of the World* (London, 2021)

McGrath, Alister, 'A consilience of equal regard: Stephen Jay Gould on the relation of science and religion', *Zygon* (2021), 56(3), 547–565

McGrath, Alister E., *The Territories of Human Reason: Science and Theology in an Age of Multiple Rationalities* (Oxford, 2019)

McGrath, Alister E., *Natural Philosophy: On Retrieving a Lost Disciplinary Imaginary* (Oxford, 2022)

McHargue, Mike, *Finding God in the Waves: How I Lost My Faith and Found It Again through Science* (London, 2016)

McLeish, Tom, *Faith and Wisdom in Science* (Oxford, 2014)

Mealand, D.L. 'The extent of the Pauline Corpus: a multivariate approach', *Journal for the Study of the New Testament* (1996), 18(59), 61–92

Merton, Robert K., 'The Normative Structure of Science', in Merton, Robert K. (ed.), *The Sociology of Science: Theoretical and Empirical Investigations* (Chicago, 1973)

Midgley, Mary, *Science and Poetry* (London, 2001)

Miller, Jon D., Scott, Eugenie C., Huffaker, Jordan S., et al., 'Public acceptance of evolution in the United States, 1985–2020', *Public Understanding of Science* (2022), 31(2), 223–238

National Academy of Sciences, *Teaching about Evolution and the Nature of Science* (Washington, DC, 1998)

Needham, Joseph, 'Human laws and laws of nature in China and the West (I)', *Journal of the History of Ideas* (1951) 12(1), 3–30

Noy, S. and O'Brien, T.L. 'A nation divided: science, religion, and public opinion in the United States'. *Socius* (2016), 2, 1–15

O'Brien, T.L. and Noy, Shiri, 'Traditional, modern, and post-secular perspectives on science and religion in the United States', *American Sociological Review* (2015), 80(1), 92–115

O'Brien, T.L. and Noy, Shiri, 'Political identity and confidence in science and religion in the United States', *Sociology of Religion: A Quarterly Review* (2020), 81(4), 439–461.

Oreskes, Naomi, *Why Trust Science?* (Princeton, NJ, 2019)

Otto, Rudolf, *The Idea of the Holy* (Oxford: OUP, 1923)

264 BIBLIOGRAPHY

Parsons, C., 'Platonism and mathematical intuition in Kurt Gödel's thought', *Bulletin of Symbolic Logic* (1995), 1(1), 44–74

Peters, Ted, 'Science and religion: ten models of war, truce, and partnership', *Theology and Science* (2018), 16, 11–53

Pigliucci, Massimo and Boudry, Maarten (eds.), *Philosophy of Pseudoscience: Reconsidering the Demarcation Problem* (Chicago, 2013)

Planck, Max, *Where Is Science Going? The Universe in the Light of Modern Physics* (London, 1933)

Plantinga, Alvin, *Where the Conflict Really Lies: Science, Religion, and Naturalism* (Oxford, 2011)

Plantinga, Alvin, and Dennett, Daniel C., *Science and Religion: Are They Compatible?* (New York, 2011)

Polkinghorne, J.C., *The Polkinghorne Reader: Science, Faith, and the Search for Meaning* (London, 2011)

Pope Benedict XVI, *Jesus of Nazareth: Part Two, from the Entrance into Jerusalem to the Resurrection* (London, 2011)

Popper, Karl, 'Philosophy of Science: A Personal Report', in Mace, C.A. (ed.) *British Philosophy in Mid-Century* (Crows Nest, New South Wales, 1957)

Ritchie, Stuart, *Science Fictions: The Epidemic of Fraud, Bias, Negligence and Hype in Science* (London, 2020)

Simmel, Georg, *Essays on Religion* (New Haven, 2013)

Sinai, Nicolai, *The Qur'an: A Historical-Critical Introduction* (Edinburgh, 2017)

Smart, Ninian, *The World's Religions* (Cambridge, 1998)

Sokal, Alan D., *The Sokal Hoax: The Sham that Shook the Academy* (Lincoln, 2000)

Sokal, Alan D. and Bricmont, Jean, *Fashionable Nonsense: Postmodern Intellectuals' Abuse of Science* (New York, 1998)

Spencer, Nick, *'Beauty Is Truth': What's Beauty Got to Do with Science?* (London, 2022)

Spencer, Nick, *Magisteria: The Entangled Histories of Science and Religion* (London, 2023)

Spencer, Nick and Waite, Hannah, *Playing God: Science, Religion and the Future of Humanity* (London, 2024)

Stamouli, Nektaria, 'Science vs. religion as Greek priests lead the anti-vax movement', *Politico* (20 July 2021)

Stenger, Victor, *God: The Failed Hypothesis: How Science Shows that God Does Not Exist* (Amherst, NY, 2007)

Taylor, Steve, 'How a flawed experiment "proved" that free will doesn't exist', *Scientific American* blog (6 December 2019)

Tillich, Paul, *Dynamics of Faith* (New York, 1958)

Torrance, Andrew, 'Should a Christian adopt methodological naturalism?', *Zygon* (September 2017), 52(3), 691–725

Ungureanu, James C., *Science, Religion, and the Protestant Tradition: Retracing the Origins of Conflict* (Pittsburgh, 2019)

Unsworth, Amy and Voas, David, 'Attitudes to evolution among Christians, Muslims and the non-religious in Britain: differential effects of religious and educational factors', *Public Understanding of Science* (2018), 27(1), 76–93

Wittgenstein, Ludwig, *Philosophical Investigations* (Oxford, 1999 [1953])

Whewell, William, *Philosophy of the Inductive Sciences* (London, 1996 [1840])

Whitehouse, Harvey, François, Pieter, Savage, Patrick E., et al., 'Retraction Note: Complex societies precede moralizing gods throughout world history', *Nature* (7 July 2021)

Wright, Robert, *Why Buddhism Is True: The Science and Philosophy of Meditation and Enlightenment* (London, 2018)

Index

A

abortion 231–42
Abrahamic religions 220
Abrahams, William 185
afterlife; *see* post-mortem existence
AIDS 38–9, 43–4, 232
alternative medicine 38, 43
altruism 205, 218, 222
American Psychiatric Association 232
Analysis of the law underpinning The Advancement of Religion for the Public Benefit 75; *see also* Charity Commission
Anglicanism 117
anthropic principle 92, 199
anthropology 17–21, 48, 82, 97, 153, 160, 186, 195–223, 252
anti-vaxxers 31, 43, 121; *see also* vaccination
Aquinas, Thomas 148
Aristotelian
    metaphysics 133
    soul, conception of 216
Arkansas 29–35
artificial contraception 231, 237
Asma, Stephen T. 10
astrology 32, 39
astronomy 19, 120, 195–6
astrophysics 19
atheism 75, 168
atheist identities 10
atheist secularists 153
atheist, evolution as 197
Atlantic, The 124

B

Babbage, Charles 33, 42
Babraham Institute 17
Baker, Fern Elsdon 197
Barbour, Ian 2, 7, 11
Barrett, Justin 212
beliefs
    philosophical 75–7

religious 94, 182
supernatural 106
system of 53, 76–84, 98, 102–3, 108, 144, 179, 211
theological 58, 187
Bellah, Robert 85
Benedict XVI, Pope 188
Bible
    Genesis 199, 214
    Hebrew 214
    literalism 34
    New Testament 158–9, 181, 215
    scholars 159
Biden administration 124
Big Bang 79, 177, 182, 197–8, 219
bioethics 228
Birmingham, University of 229
Blakely, Jason 31
Boudry, Maarten 38
British Association for the Advancement of Science 34
Brooke, John Hedley 3–4
Brown, Warren 215
Buddhism 11, 74–8, 90, 81, 107, 114, 117, 145–6, 210

C

Canterbury, Archbishop of 242–4
Cantor, Geoffrey 3
Catholic Church 202, 231–3
Catholic theologians 182
Catholicism 117, 177
ceremonies 76, 86, 99–103
Charities Act (2006) 30, 37
Charity Commission 75–7, 227
Charpentier, Emmanuelle 231
Chevassus-au-Louis, Nicolas 32, 42
Christian
    doctrine 222
    fundamentalist 202
    population of England and Wales 113
    thought and tradition 214, 232, 238

266 INDEX

Christianity 82, 106, 125, 146, 155, 168, 179, 199, 203, 236
climate change 30, 38, 44, 75
   anthropogenic 30
   denialism 38, 44
Collins, Francis 2
communism 34, 75
"Conflict and Compatibility" (article) 2
conflict by definition 107–9
consciousness
   human 203
   Krishna Consciousness Movement 79–80
   matter, as property of 153
   self-consciousness 218
conspiracy theories 38, 43
contraception, artificial 231, 237
cosmology 14, 48, 60, 97, 132, 196–201, 204, 209, 216
cover-ups, institutional 117
Covid-19 1, 122–4, 243
Cox, Brian 120
Coyne, Jerry 1, 6, 11, 41–2
Creation Research Society 34
Creation Research, Institute for 34
Creation science 29–37, 54, 144
Creation Science Research Center 34
Creationism 38, 43, 199, 213
critical analysis 15, 38, 166–70, 183–5, 189, 228
cults and sects 38

D
Darwinism 30, 34, 113, 198, 202, 211
Davis v. Beason 78, 82
Dawkins, Richard 2, 6, 41, 67
De Waal, Frans 42
demarcation problem 32–9, 43–4, 132
Dembski, William 57
"Demise of the Demarcation Problem, The" (essay) 35; *see also* Laudan, Larry
Democratic Party 228
demographics, religious 113–14
Diagnostic and Statistical Manual (*DSM*) 232
discontinuous mind 67
District Court for the Middle District of Pennsylvania 37
"Disunity of Science" (article) 38
divine action 83, 148
Divine Light Zentrum 75

doctrine
   belief in 92, 101–3
   Christian 222
   code of 80
   religious 12, 80, 197, 200–1, 204, 209, 214–16, 251–2
dogma 6, 53, 63–4, 167, 173, 178–80, 196, 203, 250
Doudna, Jennifer 231
Draper, John William 3–4
druidism 75
Drummond, Celia Deane 2
Duhem, Pierre 42
Dupré, John 38
Durkheim, Émile 84–7, 105, 144

E
ecclesiastical authority 37, 80, 188, 231
Ecklund, Elaine Howard 197–8, 203
Edgerton, David 39
Edwards v. Aguillard 20, 36–40, 50, 144, 195
Elementary Forms of Religious Life, The 84; *see also* Durkheim, Émile
Eliade, Mircea 84, 87
Elsdon-Baker, Fern 10
endorphin system 208
Enlightenment 35
environment 115, 122, 147, 214, 228
epigenetics 211
epistemic heterogeneity 35, 39
epistemic humility 52–3, 61–2, 135*t*, 168, 189, 227
Equality Act (2006) 75
essentialism 6–7, 16–18
ethical
   culture 78
   issues 142, 239–46, 251–3
   tension 236, 252
   veganism 15, 18, 75
ethics 104, 123–4, 127, 135*t*, 136–7, 175, 184, 205, 228, 234–5, 239, 244
Eucharist 133
European Convention on Human Rights 74
euthanasia 231–4, 237
evangelicalism 117
Evans, John 2, 9, 14, 20, 112–15, 127–9, 198, 228, 252
Evans, Michael 2
evolution 203–4
evolution

INDEX 267

anti-evolutionary views 251
atheistic, as 197
campaigns against 30
ethics, approach to 205
game theory 205
natural selection 182, 196, 202–4, 211, 221
psychology, evolutionary 9, 39, 206, 213–19
religion as cognitive by-product of 207
synthesis theories, extended 211
teaching of 34
theory of 68, 197, 202–5, 211, 221
existential perspectives and questions 14, 83, 97–8, 108, 134*t*
Experience of God, The 18; *see also* Hart, David Bentley
Exploring the Spectrum project 10, 229

F

faith 94–5, 101–4, 108, 114, 121, 125, 128, 165–70, 176–82, 188–9
HarrisX Global Faith and Media Study 116
institutionalized 21, 168, 189, 242
methodological shortcut, as 166–9
schools 202
falsification 50–3, 68, 108, 168–72, 183–8
Fauci, Anthony 124
Finding God in the Waves: How I Lost My Faith and Found It Again Through Science 1
Flanagan, Owen 146
Fraud in the Lab 32; *see also* Chevassus-au-Louis, Nicolas
Frazer, James 83, 87
free will 206–10, 217–19
Freeman III, George C. 81
Freudian psychoanalysis 34
fuzzy
categories 7–9, 220
definitions 49, 251
logic 38–9, 132–3

G

Geertz, Clifford 85–7, 144, 175
gender and trans rights 237
Genesis 199, 214
genetic engineering 231–4
genetically modified (GM) crops 122, 229

Gifford Lectures 73
Gingras, Yves 1–2
Glennan, Stuart 7–9
God
concept of 18, 145–6
existence of 78, 92, 195, 200, 203
Experience of God, The 18
Finding God in the Waves: How I Lost My Faith and Found It Again Through Science 1
God and Nature 3–4
God Delusion, The 41; *see also* Dawkins, Richard
God Equation, The 1
"God Helmet" 207–10
God hypothesis 199
honouring 149
interventionist, as 155–8
kingship of 238
Language of God, The: A Scientist Presents Evidence for Belief 1
law of 208
lawgiver, as 145
Mind of God, The 1
"playing God" 237–8, 241
Goldilocks hypothesis 199
Google 1
Gordin, Michael 31
Gould, Stephen Jay 2, 6, 11, 132, 141
government
ecclesiastical 80
public mistrust in 30
scientific advice 124; *see also* Covid-19
Grayling, A.C. 39
Greek New Testament 158–9, 215
Greeks, Ancient 81
Griffin, James 154

H

Hajj, the 18, 133
Harari, Yuval Noah 218
Harrison, Peter 3–18, 82, 141
HarrisX Global Faith and Media Study 116
Hart, David Bentley 18
Haught, John 2, 7
Hausser, Marc 187
Hebrew; *see* Jewish
Hindus and Hinduism 80, 113, 117, 167, 177
historical cartography 11; *see also* Harrison, Peter

268 INDEX

HIV AIDS 38–9, 43–4, 232
holy days 81, 100
Holy Spirit 182
homosexuality 116–17, 228, 232
How to Relate Science and Religion 2, 10
Howth, Transport Museum in 29–30
humility, epistemic 52–3, 61–2, 135*t*, 168, 189, 227
Humphreys, Colin 159
Hwang, Woo-Suk 187
hyperactive (or hypersensitive) agency detection device (HADD) 96, 207
hypothesis
 data and, relationship between 50–2
 God hypothesis 199
 Goldilocks hypothesis 199

I
Idea of the Holy, The 83–4; *see also* Otto, Rudolf
Immaculate Conception 119
Imperial College 39
in vitro fertilization (IVF) 231–2, 242
India, aggressive nationalism in 117
induction 37, 53, 165, 168–70, 183
Institute for Creation Research 34
intellectual cartography 5–6, 11, 141; *see also* Harrison, Peter
Intelligent Design (ID) 37–9, 44, 57, 144
International Society for Krishna Consciousness, Inc., v. Barber 79
Ipsos MORI 115
Ireland 29–30
Islam 117, 202, 214

J
James, William 73
Jesus 119, 125, 155, 158, 188, 208
Jewish
 Bible 119, 214
 God as a lawgiver, concept of 145
 Judaism 125
 population of England and Wales 114
 pork, proscription against eating 234
 tradition 214, 232, 238
Jones III, Judge John 54, 57
Jong, Jonathan 7, 11

K
Kitzmiller v. Dover Area School District 20, 39–40, 54, 57, 144, 227
knowledge, science 117–19
Krishna Consciousness Movement 79–80
Kumbh Mela 18, 133

L
Laborde, Cécile 8–15
language
 games 13–14, 250–1
 Language of God: A Scientist Presents Evidence for Belief 1
 understanding of 12–13, 17
Larson, Edward J. 3
Laudan, Larry 20, 35–8
Liberalism's Religion 8–9; *see also* Laborde, Cécile
Libet, Benjamin 210
Lightman, Bernard 3–4
Lindberg, David 3

M
Madison, James 78
magnetic resonance imaging (MRI) 210, 216–18
Malnak v. Yogi 79, 83, 144, 227
malpractice 33
Marxism 90, 34, 185
Mass, Latin 18
materialism 54–7, 102, 150–2, 196
mathematics 5, 9, 19, 29, 33, 37, 48, 56, 60, 68, 114, 133, 153–4, 160
mathematics
 Platonism 153
 statistical techniques, and 51
McDowell, John 154
McFarland, Ian 2
McGilchrist, Iain 2
McGrath, Alister 2, 7, 12
McLeish, Tom 2
measurement 37, 40, 51–2, 60, 108, 134*t*, 136, 240
media 115–17, 229
media
 narratives 229
 social 44, 119
medicine
 alternative 38, 43
 ethical 14

INDEX 269

Merton, Robert 20, 34, 39–40, 227
metaphysical
  beliefs 145, 250
  commitments 94, 142–7, 154, 160, 252
  dimension 93, 159
  naturalism 55, 235
  presuppositions 135t, 181, 195, 251
  questions 123, 145, 196
  tension 154–5, 219
metaphysics 5, 21, 52–5, 80, 93–4, 123–9,
    133–7, 142–60, 166, 181, 195–200, 204,
    219, 232, 235, 250–2
methodological dichotomy,
    re-evaluating 169–70
methodological integrity 181, 185
methodological naturalism 37, 40, 54, 57,
    144, 151–2, 160, 252
methodological science and religion, gap
    between 189
methodological theology, approach of 185
methodology 21, 166–89
  religious 177–83
  scientific 170–2
Middle Ages 5
Midgley, Mary 5–6
Mind of God, The 1
miracles 154–60
monism 56
Moon Sect 75
moral
  code 7, 79–81
  concepts 126, 205
  judgement 92
  realism 153, 217
  sensitivity 218
morality 3, 82, 97–9, 108, 124–7, 185,
    203–9, 217–21, 232–6
Morals Not Knowledge: Recasting the
    Contemporary U.S. Conflict between
    Religion and Science 9, 112; see also
    Evans, John
multiverse 154, 199–200
Murdoch, Iris 154
Murphy, Nancy 215
Muslim population of England and
    Wales 113
Myanmar, mistreatment of minorities in 117
mysterious, the 87, 97, 100, 160
mythic dimension 86–8
mythical

age 84
Creation stories 83
depiction of science 42, 59
mythologies 96
texts 167

N
nanotechnology 122, 229
National Academy of Sciences (NAS) 54
natural law 35, 40, 58, 144, 158
natural phenomena 36, 40, 83, 144
natural selection 182, 196, 202–4, 211, 221
naturalism
  liberal 55, 151
  metaphysical 55, 235
  methodological 37, 40, 54, 57, 144, 151–2,
    160, 252
  ontological 55, 151, 160, 252
  scientific 9, 151
Nature of Religion, The 84; see also Mircea
    Eliade
Nature journal 1
nature, laws of 158–9
Nazism 75
Needham, Joseph 145
neuroscience 3, 32, 196, 204–9, 215–19
New Testament
  Greek 158–9, 215
  miracles, concept of 158
  Pauline corpus in 181
Newton, Sir Isaac 55, 150, 154
niche construction theory 211
non-overlapping magisteria (NOMA),
    principle of 2, 6
Noy, Shiri 228
Numbers, Ronald 3–4
numinous 83–7

O
O'Brien, Timothy 228
Office of Science and Technology Policy 33
oil industry 30
ontological naturalism 160
ordered observation 51–2
Oreskes, Naomi 30, 174
Organisation for Economic Cooperation
    and Development (OECD) 33
organized scepticism 21, 34, 40

270 INDEX

Otto, Rudolph 83–7
Overton Judgement 20, 30, 36–40, 144, 171, 227

P

pacifism 75
Paley, William 203, 221
panpsychism 153
parapsychology 32
PAS survey 115–24
Pauline corpus in the New Testament 181
Persinger, Michael 207–10
Peters, Ted 2, 7
phenomena
  natural 36, 40, 83, 144
  non-natural 149
philosophers 14, 19, 34, 43, 47, 54–6, 87, 97, 101, 106, 128, 145–6, 151–4, 170–1, 175, 180, 203, 218, 253
philosophical belief 75–7
Philosophical Investigations 12; *see also* Wittgenstein, Ludwig
Philosophy of the Inductive Sciences 34
physicalism 56, 150–4
physics 5, 19, 29–33, 38–9, 47, 53–6, 60–1, 68, 120, 184, 195
Pigliucci, Massimo 20, 38–9, 48
Pinker, Steven 6
Plantinga, Alvin 57
Platonism, mathematical 153
Platonists 56, 82
Political Liberalism 245; *see also* Rawls, John
political theory 8, 184
Polkinghorne, John 2, 183, 216
Popper, Karl 20, 34–5, 50, 171–2
post-mortem existence 196, 208, 220
Princeton University 174
progressive v. conservative 234–5
pro-life 75
Proof of Heaven: A Neurosurgeon's Journey into the Afterlife 1
Protestant, -ism
  charismatic 179
  Christianity 82–3, 90, 177–9
  fundamentalism 177
proto- or quasiscience 39, 127, 207, 234
pseudoscience 31–4, 43, 38–9
psychic powers 75
psychoanalysis 17, 34, 38
psychology, evolutionary 9, 39, 206, 213–19

Public Attitudes to Science (PAS) 115–24
public authority 21, 227–47, 252–3
public mistrust in government 30
public opinion 21, 34, 113, 121, 126, 198, 202, 229, 243
public reasoning 227–47, 252–3
public trust in science and scientists 31, 122–4, 176

Q

Quakers 83
quantification 51–3, 68, 134*t*, 136, 168, 181–4
quantum
  fluctuations 199–200
  gravity 31
  mechanics 154, 182, 200
  theory 68, 132, 151

R

R (on the Application of Hodkin) v. Registrar General of Births, Deaths and Marriages 76–7, 144
R v. Registrar General, ex parte Segerdal 74–7
race 228
radicalization 4
Rawls, John 8, 245
Re South Place Ethical Society, Barralet v. AG 74–7
Reflections on the Decline of Science in England 33; *see also* Babbage, Charles
religion
  Abrahamic 220
  anthropology of 82
  belief system, as 53, 144
  boundary conditions 90, 105
  conception of 181
  cultural understanding of 126–7
  defining 20, 73, 77–82, 87–8, 125
  dimensions of 8, 11, 15, 101, 144, 176
  disaggregating 90–109
  evolution, as cognitive by-product of 207
  faith or spirituality, juxtaposing with 104
  faith, as formalization of 102
  family definition of 91, 105, 126, 142, 165
  "hardness" of 54
  institutionalization of 88, 100, 104–6, 135t, 238, 251
  philosophy of 73, 82

polysemy of 82–7
protoscience, as 127, 207
public understanding of 124–9
ritual understanding of 146–7, 180, 196
science and, conflict between 3, 21, 195,
  220, 230, 242
science and, landscapes and territories
  of 3–7, 16–18, 160, 166–70, 196–8, 219,
  227–8, 233, 244, 247
science and, public understanding
  of 113–20
social or communal endeavour, as 88
soul, ideas about 186, 204–9, 215–20, 232
supernatural element in 85
supernaturalism, commitment to 147–50
symbols, as system of 144; see also
  Geertz, Clifford
transcendent dimension to 92–3
unified system of beliefs and practices,
  as 144; see also Durkheim, Émile
religious
  belief 94–8, 182, 101–3
  commitment 8, 98–101, 147–50, 216
  demographics 113–14
  doctrine 12, 80, 197, 200–1, 204, 209,
    214–16, 251–2
  experience 167, 181–5, 210
  hypocrisy 238
  methodology 177
  protocols 105
  publics 230–3
  revelation 37, 165–8, 178–86, 189, 244,
    252
  rites 76
  rituals 84, 100–6, 129, 135t, 136, 179–80,
    208
  scientific disciplines and theories that
    make it hard to be religious 196–7
  services 80, 230–3
  studies 19, 85–7, 115, 155, 178–80
  texts 8, 114, 134t, 135t, 136, 146, 159,
    167–8, 181–6, 199, 245, 251
repetition 52, 165, 171, 189
replication 32–3, 42–4, 52
Republican Party 228
Rethinking History, Science and Religion 4;
  see also Lightman, Bernie
revelation, divine or religious 37, 165–8,
  178–86, 189, 244, 252
Ritchie, Stuart 32, 42

rites 76, 100
ritual, -s
  commitment 99–102
  culture 88, 126–7
  laws, and offices, formalization of belief
    in 91, 103–5
  nature, as attempt to control 83
  practice 100
  Quaker distrust of 83
  religion, ritual understanding of 146–7,
    180, 196
  religious 84, 100–6, 129, 135t, 136,
    179–80, 208
  sacred 81
  secular 106

S
sacred, the 81–7
sacred
  life as 245
  rituals 81
  soul as 232
  texts, religious 8, 114, 134t, 135t, 136,
    146, 159, 167–8, 181–6, 199, 245, 251
  things 84–5
  time 84
scepticism, organized 21, 34, 40–1, 121
Schon, Jan Hendrik 187
Science and Religion: New Historical
    Perspectives 4
Science and Religion: Some Historical
    Perspectives 3–4
"Science and Technology in a Democratic
    Order" (essay) 34; see also Merton,
    Robert
Science Council 17, 20, 37–40, 144, 165, 227
Science Fictions 32; see also Ritchie, Stuart
science
  confidence and science knowledge, cross
    tabulation of 118t
  "Creation science" 29
  defining 17, 29–44
  disaggregated features of 48
  "disbelief system", as 53
  experimentation as central to 51–3, 68,
    168, 183
  falsification central feature in 68
  "family definition" of 20–1, 39–41, 49–50,
    59–67, 73, 142, 165, 195, 227
  institution, as 43, 64–6, 175–6

272 INDEX

science (*Continued*)
  institutionalization of 65–6, 103, 176, 251
  knowledge 117–19
  methods of 49–54
  mythical depiction of 42, 59
  naturalism, commitment to 150–4
  objectives of 59–61
  practice of 173–7
  public conceptions of 21
  public trust in 31, 122–4, 176
  public understanding of 113–24
  publics 229–33
  refutation 51–3, 172
  religion and, conflict between 3, 21, 195,
    220, 230, 242
  religion and, disaggregating 8–16, 47–69,
    112–29
  religion and, landscapes and territories
    of 3–7, 16–18, 160, 166, 169–70, 196–8,
    219, 227–8, 233, 244, 247
  replicability 32, 51
  science-in-theory 169, 174
  science–religion debate 60, 113
  self-refutation 172
  subject of 54–7
  *Teaching about Evolution and the Nature
    of Science* 54
  testability 37, 50, 53, 68
  values of 61–4
Scientific Advisory Group for Emergencies
  (SAGE) 243
scientific
  curiosity 61–3, 135*t*
  disciplines 47, 51, 60–1, 68, 151, 187,
    196–7, 200, 204, 213–19
  falsification 171–2, 183
  integrity, breaches in 33
  methodology 171–2
  naturalism 9, 151
scientism 77, 150–1
Scientology 38, 74–6
Search for Extraterrestrial Intelligence 39
Secular Humanism 75, 78
secularists, atheist 153
secularization 4
Serpent's Promise, The 1
sex abuse 117
sexuality 228, 232
Sikhism 114, 125
Simmel, George 82

"Simplifying Complexity" (essay) 4; *see also*
  Numbers, Ronald
Smart, Ninian 11, 85–7, 144, 176, 227
social
  issues 252
  media 44, 119
sociology 2, 19, 39, 48, 64, 82
"soft" sciences 39
Sokal, Alan 31
Sorites paradox 38
soul
  Aristotelian conception of 216
  belief in 83, 129
  ideas about 186, 204–9, 215–20, 232
spiritual
  belief 1, 3, 76
  disciplines 135*t*, 200
  experience 181–2, 187, 219
  improvement 76
  issues 126–9
  practices 105
  principle 75
  spiritualism 75
  spirituality 81, 90, 101–4
  transcendence, language of 92
  truths 160
  understanding 77
Stapel, Diederik 187
State Academies of Science 36
stem-cell research 122, 229, 232
Stenger, Victor 2
Stenmark, Mikael 2, 10
string theory 48, 154, 172
Sufis 167
suicide, physician-assisted 231–2, 242
Sunday schools 80
supernatural 55, 84–5, 93, 146–50, 154, 158
  agency 7, 92–4, 107, 134*t*
  beings and entities 15, 85, 92–4
  beliefs 106
  causation 54
  connection with 92–3
  element in religion 85, 92, 96, 146
  explanations 36, 155
  intervention 96
  religion's commitment to 147–50
  supernaturalism 21, 145–55, 160
Supreme and Circuit Court definitions of
  religion 81
supreme being, belief in 75, 81, 195
Supreme Court 36–7, 74, 78–9, 82

INDEX 273

T

Tanner Lectures on Human Values
  published as *Why Trust Science?* 30,
  174; *see also* Oreskes, Naomi
Taoism 78
Taylor, Charles 87
*Teaching about Evolution and the Nature of
  Science* 54
Templeton Religion Trust 19
Territories of Human Reason 12; *see also*
  McGrath, Alister
territories of science and religion 3–8; *see
  also* Harrison, Peter
terrorism 117
Tertullian 166
texts, religious 8, 114, 134*t*, 135*t*, 136, 146,
  159, 167–8, 181–6, 199, 245, 251
theology 20, 48, 54, 91, 101–3, 135*t*, 166,
  178–89, 201, 211, 221–2
  beliefs, theological 20, 58, 91–2, 101–3,
    187
theory
  evolutionary 68, 197, 202–5, 211, 221
  Intelligent Design (ID) 37–9, 44, 57, 144
  language, of 12
  niche construction 211
  political 8, 184
  quantum 132, 151
  science-in-theory 169, 174
  string 48, 154, 172
  Theory of Justice 245; *see also* Rawls, John
  universe's origins, on 199
Theos/Faraday research 114, 118, 121–4,
  196–7
Tillich, Paul 83–4, 87, 144, 227
Toulson, Lord 74–9, 147
trans rights 237
transcendence 20, 81, 91–4, 105, 135*t*, 207
transcendent, the 99, 104–6, 141, 149, 152,
  196, 209, 251–2
  being, transcendental 87, 212
  reality, transcendent dimension to 134*t*,
    149
  religion, transcendent dimension to 92–3
  world, transcendent engagement
    with 212
Transport Museum in Howth 29–30
Trump, Donald 117
trust, public

government, mistrust in 30
science and scientists, in 31, 122–4, 176
Tylor, E.B. 83

U

UFOlogy 32
Ungureanu, James 3
United States v. Seeger 78–9, 83
universe, mysterious 97, 100
Unsworth, Amy 197

V

vaccination 1, 30–2, 122
  anti-vaxxers 31, 43, 121
  Covid-19 1, 124
Values, Human, Tanner Lectures on 174
Vatican 187
veganism, ethical 15, 18, 75
verification 38, 165, 196
Virginia Polytechnic Institute and State
  University 32

W

Weinberg, Steven 1
When Science Meets Religion 2; *see also*
  Barbour, Ian
*Where the Conflict Really Lies: Science,
  Religion and Naturalism* 9; *see also*
  Plantinga, Alvin
Whewell, William 34
White, Andrew Dickson 3–4
"Whose science and whose religion?"
  (article) 7; *see also* Glennan, Stuart
Why Trust Science? 30, 174; *see also*
  Oreskes, Naomi
Wiggins, David 154
Wittgenstein, Ludwig 11–20, 38–9, 67–8, 81,
  91
  "family resemblance" approach 20
  language as usage 17
  meaning as usage 14
  Philosophical Investigations 12
worship
  creed and form of 80
  prayer and 81
Wright, Robert 146

Y

YouGov 114, 124
Youmans, Edward 3